Mastering Android NDK

Master the skills you need to develop portable, highly-functional Android applications using NDK

Sergey Kosarevsky

Viktor Latypov

PUBLISHING

BIRMINGHAM - MUMBAI

Mastering Android NDK

First published: September 2015

Production reference: 2161015

Published by Packt Publishing Ltd.
Livery Place
35 Livery Street
Birmingham B3 2PB, UK.

ISBN 978-1-78528-833-3

www.packtpub.com

Credits

Authors
Sergey Kosarevsky
Viktor Latypov

Reviewers
Koji Hasegawa
Raimon Ràfols
Sebastian Roth
Ali Utku Selen

Commissioning Editor
Priya Singh

Acquisition Editor
Tushar Gupta

Content Development Editor
Neeshma Ramakrishnan

Technical Editor
Dhiraj Chandanshive

Copy Editors
Janbal Dharmaraj
Kevin McGowan

Project Coordinator
Shweta H Birwatkar

Proofreader
Safis Editing

Indexer
Hemangini Bari

Production Coordinator
Nitesh Thakur

Cover Work
Nitesh Thakur

About the Authors

Sergey Kosarevsky is a software engineer with experience in C++ and 3D graphics. He worked for mobile industry companies and was involved in mobile projects at SPB Software, Yandex, Layar and Blippar. He has more than 12 years of software development experience and more than 6 years of Android NDK experience. Sergey got his PhD in the field of mechanical engineering from St. Petersburg Institute of Machine-Building in Saint-Petersburg, Russia. In his spare time, Sergey maintains and develops an open source multiplatform gaming engine, the Linderdaum Engine (`http://linderdaum.com`).

> I would like to thank my mother, Irina Kosarevskaya, and my grandma, Ludmila Sirotkina. Without their unconditional love and support, all of this could never have happened.

Viktor Latypov is a software engineer and mathematician with experience in compiler development, device drivers, robotics, and high-performance computing and with a personal interest in 3D graphics and mobile technology. Surrounded by computers for more than 20 years, he enjoys every bit of developing and designing software for anything with a CPU inside. Viktor holds a PhD in applied mathematics from Saint-Petersburg State University.

About the Reviewers

Koji Hasegawa is an iOS/Android app developer, living in Japan. He released various applications to the AppStore, such as "Den-Ace quiz, Minoru Kawasaki's tokusatsu movies world", and "Futsal rules and trivia".

In addition, he is also involved in community activities related to test automation. Koji has authored a few books on testing and test automation in Japanese.

He is an ISTQB Certification Tester and is also certified in Iaido (Japanese Sword Arts) and Kyudo (Japanese Archery).

Raimon Ràfols has been developing for mobile devices since 2004. He has experience in developing on several technologies specializing in the UI, build systems, and client-server communications. He is currently working as an engineering manager at AXA Group Solutions in Barcelona. In the past, he worked for Imagination Technologies near London and Service2Media in the Netherlands. In his spare time, he enjoys programming, photography, and giving talks on Android performance optimization and Android custom views at mobile conferences.

Ali Utku Selen is a system engineer at Sony Mobile Communications, working on flagship Android devices for more than 5 years. He started programming at the age of 11, and since then has had a great interest in software development. He also holds an MSc degree from Dokuz Eylül University, Computer Engineering Department.

www.PacktPub.com

Support files, eBooks, discount offers, and more

For support files and downloads related to your book, please visit www.PacktPub.com.

Did you know that Packt offers eBook versions of every book published, with PDF and ePub files available? You can upgrade to the eBook version at www.PacktPub.com and as a print book customer, you are entitled to a discount on the eBook copy. Get in touch with us at service@packtpub.com for more details.

At www.PacktPub.com, you can also read a collection of free technical articles, sign up for a range of free newsletters and receive exclusive discounts and offers on Packt books and eBooks.

https://www2.packtpub.com/books/subscription/packtlib

Do you need instant solutions to your IT questions? PacktLib is Packt's online digital book library. Here, you can search, access, and read Packt's entire library of books.

Why subscribe?

- Fully searchable across every book published by Packt
- Copy and paste, print, and bookmark content
- On demand and accessible via a web browser

Free access for Packt account holders

If you have an account with Packt at www.PacktPub.com, you can use this to access PacktLib today and view 9 entirely free books. Simply use your login credentials for immediate access.

Dedicated to my grandmother Ludmila Fedorovna Sitorkina and my mother Irina Leonidovna Kosarevskaya.

– Sergey Kosarevsky

Dedicated to my family: parents, wife and numerous nephews.

– Viktor Latypov

Table of Contents

Preface

This book is a sequel to *Android NDK Game Development Cookbook, Packt Publishing*, published in 2013. It covers NDK development from quite an unusual point of view: building your mobile C++ applications in a portable way so that they can be developed and debugged on a desktop computer. This approach greatly reduces iteration and content integration time and is essential in the world of professional mobile software development.

What this book covers

Chapter 1, Using Command-line Tools, shows you how to install and configure the essential tools for Android native development using the command line and how to write basic Android application configuration files manually from scratch without falling back on the graphical IDEs.

Chapter 2, Native Libraries, shows you how to build popular C/C++ libraries and link them against your applications using the Android NDK. These libraries are the building blocks to implement feature-rich applications with images, videos, sounds, and networking entirely in C++. We will show you how to compile libraries and, of course, give some examples and hints on how to start using them. Some of the libraries are discussed in greater detail in the subsequent chapters.

Chapter 3, Networking, focuses on how to deal with network-related functionality from the native C/C++ code. Networking tasks are asynchronous by nature and unpredictable in terms of timing. Even when the underlying connection is established using the TCP protocol, there is no guarantee on the delivery time, and nothing prevents the applications from freezing while waiting for the data. We will take a closer look at implementing basic asynchronous mechanisms in a portable way.

Chapter 4, Organizing a Virtual Filesystem, implements low-level abstractions to deal with the OS-independent access to files and filesystems. We will show how to implement portable and transparent access to the Android assets packed inside the .apk files without falling back on any built-in APIs. This approach is necessary when building multiplatform applications that are debuggable in a desktop environment.

Chapter 5, Cross-platform Audio Streaming, implements a truly portable audio subsystem for Android and desktop PCs based on the OpenAL library. The code makes use of the multithreading material from *Chapter 3, Networking*.

Chapter 6, OpenGL ES 3.1 and Cross-platform Rendering, focuses on how to implement an abstraction layer on top of OpenGL 4 and OpenGL ES 3 to make our C++ graphics applications runnable on Android and desktop machines.

Chapter 7, Cross-platform UI and Input System, details the description of a mechanism to render geometric primitives and Unicode text. The second part of the chapter describes a multi-page graphical user interface suitable for being the cornerstone for building the interfaces of multiplatform applications. This chapter concludes with an SDL application, which demonstrates the capabilities of our UI system in action.

Chapter 8, Writing a Rendering Engine, will move you into the actual rendering territory and use the thin abstraction layer, which is discussed in *Chapter 6, OpenGL ES 3.1 and Cross-platform Rendering*, to implement a 3D rendering framework capable of rendering geometry loaded from files using materials, lights, and shadows.

Chapter 9, Implementing Game Logic, introduces a common approach to organize interactions between gaming code and the user interface part of the application. The chapter begins with an implementation of the Boids algorithm and then proceeds with the extension of our user interface implemented in the previous chapters.

Chapter 10, Writing Asteroids Game, continues to put together the material from previous chapters. We will implement an Asteroids game with 3D graphics, shadows, particles, and sounds using techniques and code fragments introduced in the previous chapters.

What you need for this book

This book assumes that you have a Windows-based PC. An Android smartphone or tablet is advisable due to the limitations of the emulator in 3D graphics and native audio.

The source code in this book is based on open source Linderdaum Engine and is a hard squeezing of some approaches and techniques used in the engine. You can get it at http://www.linderdaum.com.

Basic knowledge of C or C++, including pointer manipulation, multithreading, and basic object-oriented programming concepts is assumed. You should be familiar with advanced programming concepts such as threading and synchronization primitives, and have some basic understanding of GCC toolchains. Android Java development is not covered in this book. You will have to read something else to get familiar with it.

Some working knowledge of linear algebra and affine transformations in 3D space is useful for the understanding of 3D graphics-related chapters.

Who this book is for

This book is intended for existing Android developers who are familiar with the fundamentals of the Android NDK and wish to gain expertise in game development using the Android NDK. Readers must have reasonable experience of Android application development.

Conventions

In this book, you will find a number of styles of text that distinguish between different kinds of information. Here are some examples of these styles, and an explanation of their meaning.

Code words in text, database table names, folder names, filenames, file extensions, pathnames, dummy URLs, user input, and Twitter handles are shown as follows: "Compilation of an Android static library requires a usual set of Android.mk and Application.mk files."

A block of code is set as follows:

```
std::string ExtractExtension( const std::string& FileName )
{
  size_t pos = FileName.find_last_of( '.' );
  return ( pos == std::string::npos ) ?
    FileName : FileName.substr( pos );
}
```

When we wish to draw your attention to a particular part of a code block, the relevant lines or items are set in bold:

```
std::string ExtractExtension( const std::string& FileName )
{
  size_t pos = FileName.find_last_of( '.' );
  return ( pos == std::string::npos ) ?
    FileName : FileName.substr( pos );
}
```

Any command-line input or output is written as follows:

```
>ndk-build
>ant debug
>adb install -r bin/App1-debug.apk
```

New terms and **important words** are shown in bold. Words that you see on the screen, in menus or dialog boxes for example, appear in the text like this: "Check the line **Hello Android NDK!** printed into the Android system log."

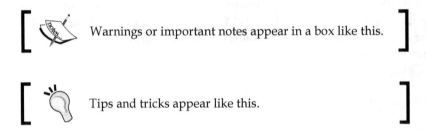

Warnings or important notes appear in a box like this.

Tips and tricks appear like this.

Reader feedback

Feedback from our readers is always welcome. Let us know what you think about this book—what you liked or may have disliked. Reader feedback is important for us to develop titles that you really get the most out of.

To send us general feedback, simply send an e-mail to feedback@packtpub.com, and mention the book title via the subject of your message.

If there is a topic that you have expertise in and you are interested in either writing or contributing to a book, see our author guide on www.packtpub.com/authors.

Customer support

Now that you are the proud owner of a Packt book, we have a number of things to help you to get the most from your purchase.

Downloading the example code

You can download the example code files for all Packt books you have purchased from your account at http://www.packtpub.com. If you purchased this book elsewhere, you can visit http://www.packtpub.com/support and register to have the files e-mailed directly to you. The source code is also available as a GitHub repository from this URL https://github.com/corporateshark/Mastering-Android-NDK. Check it to get the most up-to-date version of the sources.

Errata

Although we have taken every care to ensure the accuracy of our content, mistakes do happen. If you find a mistake in one of our books—maybe a mistake in the text or the code—we would be grateful if you would report this to us. By doing so, you can save other readers from frustration and help us improve subsequent versions of this book. If you find any errata, please report them by visiting http://www.packtpub.com/submit-errata, selecting your book, clicking on the **errata submission form** link, and entering the details of your errata. Once your errata are verified, your submission will be accepted and the errata will be uploaded on our website, or added to any list of existing errata, under the Errata section of that title. Any existing errata can be viewed by selecting your title from http://www.packtpub.com/support.

Piracy

Piracy of copyright material on the Internet is an ongoing problem across all media. At Packt, we take the protection of our copyright and licenses very seriously. If you come across any illegal copies of our works, in any form, on the Internet, please provide us with the location address or website name immediately so that we can pursue a remedy.

Please contact us at copyright@packtpub.com with a link to the suspected pirated material.

We appreciate your help in protecting our authors, and our ability to bring you valuable content.

Questions

You can contact us at questions@packtpub.com if you are having a problem with any aspect of the book, and we will do our best to address it.

1
Using Command-line Tools

In this chapter, we will take a tour of the main command-line tools, specifically related to the creation and packaging of Android applications. We will learn how to install and configure Android NDK on Microsoft Windows, Apple OS X, and Ubuntu/Debian Linux, and how to build and run your first native application on an Android-based device. Usage of command-line tools to build your projects is essential for cross-platform mobile development using C++.

 This book is based on the Android SDK revision 24.3.3 and the Android NDK r10e. The source code was tested with Android API Level 23 (Marshmallow).

Our main focus will be the command-line centric and platform-independent development process.

 Android Studio is a very nice new portable development IDE, which has recently arrived at version 1.4. However, it still has very limited NDK support and will not be discussed in this book.

Using Android command-line tools on Windows

To start developing native C++ applications for Android in a Microsoft Windows environment, you will need some essential tools to be installed on your system.

Start NDK development for Android using the following list of all the prerequisites you will need:

- The Android SDK: You can find this at `http://developer.android.com/sdk/index.html`. We use revision 24.

- The Android NDK: You can find this at `http://developer.android.com/tools/sdk/ndk/index.html`. We use version r10e.

- The **Java Development Kit (JDK)**: You can find this at `http://www.oracle.com/technetwork/java/javase/downloads/index.html`. We use Oracle JDK Version 8.

- Apache Ant: You can find this at `http://ant.apache.org`. This is a tool used to build Java applications.

- Gradle: You can find this at `https://www.gradle.org`. Compared to Ant, this is a more modern Java build automation tool capable of managing external dependencies.

The current versions of these tools will run on Windows without using any intermediate compatibility layer; they do not require Cygwin any more.

As much as it pains us to write this, Android SDK and NDK should still be installed into folders that do not contain any whitespaces in their names. This is a limitation of build scripts within the Android SDK; the unquoted environment variables content are split into words based on tab, space and newline characters.

We will install the Android SDK to `D:\android-sdk-windows`, the Android NDK to `D:\ndk`, and other software to their default locations.

In order to compile our portable C++ code for Windows, we need a decent toolchain. We recommend using the latest version of the MinGW from the Equation package available at `http://www.equation.com`. You can choose 32- or 64-bit versions as you go.

Once all the tools are in their folders, you need to set environment variables to point to those install locations. The `JAVA_HOME` variable should point to the Java Development Kit folder:

```
JAVA_HOME="D:\Program Files\Java\jdk1.8.0_25"
```

The NDK_HOME variable should point to the Android NDK installation folder:

`NDK_HOME=D:\NDK`

The ANDROID_HOME should point to the Android SDK folder:

`ANDROID_HOME=D:\\android-sdk-windows`

 [Note the double backslash in the last line.]

Both NDK and SDK will have new versions from time to time, so it might be helpful to have the version number on the folder name and manage NDK folders per project if necessary.

Using Android command-line tools on OS X

Installation of Android development tools on OS X is straightforward. First of all, you will need to download the required official SDK and NDK packages from `http://developer.android.com/sdk/index.html`. As we are going for command-line tools, we can use the SDK Tools Only package available at `http://dl.google.com/android/android-sdk_r24.0.2-macosx.zip`. As for the NDK, OS X Yosemite works with the 64-bit Android NDK, which can be downloaded from `http://developer.android.com/tools/sdk/ndk/index.html`.

We will install all these tools into the user's home folder; in our case, it is `/Users/sk`.

To get Apache Ant and Gradle, the best way would be to install the package manager Homebrew from `http://brew.sh` and bring in the required tools using the following commands:

```
$ brew install ant
$ brew install gradle
```

This way you will not be bothered with installation paths and other low-level configuration stuff. The following are the steps to install packages and set path for them:

 Since the notion of this book is doing stuff from the command line, we will indeed do so the hard way. However, you are encouraged to actually visit the download page, http://developer.android. com/sdk/index.html, in your browser and check for updated versions of the Android SDK and NDK.

1. Download the Android SDK for OS X from the official page and put it into your home directory:

    ```
    >curl -o android-sdk-macosx.zip
    http://dl.google.com/android/android-sdk_r24.0.2-macosx.zip
    ```

2. Unpack it:

    ```
    >unzip android-sdk-macosx.zip
    ```

3. Then, download the Android NDK. It comes as a self-extracting binary:

    ```
    >curl -o android-ndk-r10e.bin
    http://dl.google.com/android/ndk/android-ndk-r10e-darwin-
    x86_64.bin
    ```

4. So, just make it executable and run it:

    ```
    >chmod +x android-ndk-r10e.bin

    >./android-ndk-r10e.bin
    ```

5. The packages are in place. Now, add paths to your tools and all the necessary environment variables to the .profile file in your home directory:

    ```
    export PATH=/Users/sk/android-ndk-r10e:/Users/sk/android-ndk-
    r10e/prebuilt/darwin-x86_64/bin:/Users/sk/android-sdk-
    macosx/platform-tools:$PATH
    ```

6. Use these variables within Android scripts and tools:

    ```
    export NDK_ROOT="/Users/sk/android-ndk-r10e"

    export ANDROID_SDK_ROOT="/Users/sk/android-sdk-macosx"
    ```

7. Edit the local.properties file to set up the paths on a per-project basis.

Using Android command-line tools on Linux

Installation on Linux is as easy as its OS X counterpart.

 Indeed, Linux development environment is truly native for all kinds of Android development since all the toolchains and Android Open Source Project are based on Linux tools.

Here, we will point out just some differences. First of all, we don't need to install Homebrew. Just go with the available package manager. On Ubuntu, we prefer using `apt`. The following are the steps to install the packages as well as set path on Linux:

1. Let's start with updating all `apt` packages and installing the default Java Development Kit:

   ```
   $ sudo apt-get update
   $ sudo apt-get install default-jdk
   ```

2. Install the Apache Ant build automation tool:

   ```
   $ sudo apt-get install ant
   ```

3. Install Gradle:

   ```
   $ sudo apt-get install gradle
   ```

4. Download the official Android SDK which suits your version of Linux from `http://developer.android.com/sdk/index.html`, and unpack it into a folder in your home directory:

   ```
   $ wget http://dl.google.com/android/android-sdk_r24.0.2-linux.tgz

   $ tar -xvf android-sdk_r24.0.2-linux.tgz
   ```

5. Download the official NDK package suitable for your Linux, 32- or 64-bit, and run it:

   ```
   $ wget http://dl.google.com/android/ndk/android-ndk-r10e-linux-x86_64.bin

   $ chmod +x android-ndk-r10e-linux-x86_64.bin

   $ ./android-ndk-r10e-linux-x86_64.bin
   ```

   The executable will unpack the content of the NDK package into the current directory.

6. Now you can set up the environment variables to point to the actual folders:

```
NDK_ROOT=/path/to/ndk
```

```
ANDROID_HOME=/path/to/sdk
```

 It is useful to add environment variables definitions to /etc/profile or /etc/environment. This way these settings will be applicable to all the users of the system.

Creating an Ant-based application template manually

Let's start with the lowest level and create a template for our applications buildable with Apache Ant. Every Android application which is to be built using Apache Ant should contain a predefined directories structure and configuration .xml files. This is usually done using Android SDK tools and IDEs. We will explain how to do it by hand to let you know the machinery behind the curtains.

 Downloading the example code

You can download the example code files from your account at http://www.packtpub.com for all the Packt Publishing books you have purchased. If you purchased this book elsewhere, you can visit http://www.packtpub.com/support and register to have the files e-mailed directly to you.

For this book, the source code files can be downloaded or forked from the following GitHub repository as well: https://github.com/corporateshark/Mastering-Android-NDK

The directory structure of our minimalistic project looks like the following screenshot (see the source code bundle for the complete source code):

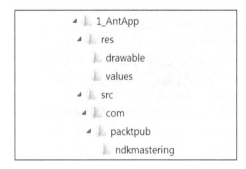

We need to create the following files within this directory structure:

- `res/drawable/icon.png`
- `res/values/strings.xml`
- `src/com/packtpub/ndkmastering/App1Activity.java`
- `AndroidManifest.xml`
- `build.xml`
- `project.properties`

The icon `icon.png` should be there, and currently contains a dummy image of an Android application:

The file `strings.xml` is required to make use of the Android localization system. In the manifest `AndroidManifest.xml`, we use the string parameter `app_name` instead of the actual application name. The file `strings.xml` resolves this parameter into a human readable string:

```xml
<?xml version="1.0" encoding="utf-8"?>
<resources>
  <string name="app_name">AntApp1</string>
</resources>
```

The Java source code of the minimal buildable application is in the `App1Activity.java` file:

```java
package com.packtpub.ndkmastering;
import android.app.Activity;
public class App1Activity extends Activity
{
};
```

The rest three files, `AndroidManifest.xml`, `build.xml`, and `project.properties`, contain the description of the project necessary for Ant to build it.

The manifest `AndroidManifest.xml` is as follows:

```
<?xml version="1.0" encoding="utf-8"?>
<manifest
xmlns:android="http://schemas.android.com/apk/res/android"
package="com.packtpub.ndkmastering"
android:versionCode="1"
android:versionName="1.0.0">
```

Our application will require Android 4.4 (API Level 19) and is tested with Android 6.0 (API Level 23):

```
<uses-sdk android:minSdkVersion="19" android:targetSdkVersion="23"
/>
```

Most of the examples in this book will require OpenGL ES 3. Let's mention it here:

```
<uses-feature android:glEsVersion="0x00030000"/>
<application android:label="@string/app_name"
android:icon="@drawable/icon"
android:installLocation="preferExternal"
android:largeHeap="true"
android:allowBackup="true">
```

Here is the name of the main activity:

```
<activity android:name="com.packtpub.ndkmastering.App1Activity"
android:launchMode="singleTask"
```

We want a fullscreen application in the landscape orientation:

```
android:theme="@android:style/Theme.NoTitleBar.Fullscreen"
android:screenOrientation="landscape"
```

Our application can be started from the system launcher. The displayable name of the application is stored in the `app_name` parameter:

```
android:configChanges="orientation|keyboardHidden"
android:label="@string/app_name">
<intent-filter>
  <action android:name="android.intent.action.MAIN" />
  <category android:name="android.intent.category.LAUNCHER" />
</intent-filter>
</activity>
</application>
</manifest>
```

You can read the official Google documentation on the application manifest at `http://developer.android.com/guide/topics/manifest/manifest-intro.html`.

The file `build.xml` is much simpler and will resemble mostly what Android tools would generate:

```xml
<?xml version="1.0" encoding="UTF-8"?>
<project name="App1" default="help">
  <loadproperties srcFile="project.properties" />
  <fail message="sdk.dir is missing. Make sure to generate
    local.properties using 'android update project' or to inject it
    through an env var"
    unless="sdk.dir"/>
  <import file="${sdk.dir}/tools/ant/build.xml" />
</project>
```

There is a difference to Android SDK Tools, since we don't use `ant.properties` here. This was done just for the sake of simplicity and just has an educational purpose.

The same situation exists with the file `project.properties`, which contains platform-specific declarations:

target=android-19

sdk.dir=d:/android-sdk-windows

Now, our first application (which does not even contain any native code yet) is ready to be built. Use the following one-liner to build it:

$ ant debug

If everything was done correctly, you should see the tail of the output similar to the following:

```
-post-package:

-do-debug:
 [zipalign] Running zip align on final apk...
    [echo] Debug Package: F:\Book_MasteringNDK\Sources\Chapter1\1_AntApp\bin\App1-debug.apk
[propertyfile] Creating new property file: F:\Book_MasteringNDK\Sources\Chapter1\1_AntApp\bin\build.prop
[propertyfile] Updating property file: F:\Book_MasteringNDK\Sources\Chapter1\1_AntApp\bin\build.prop
[propertyfile] Updating property file: F:\Book_MasteringNDK\Sources\Chapter1\1_AntApp\bin\build.prop
[propertyfile] Updating property file: F:\Book_MasteringNDK\Sources\Chapter1\1_AntApp\bin\build.prop

-post-build:

debug:

BUILD SUCCESSFUL
Total time: 2 seconds
```

To install an `.apk` file from the command line, run `adb install -r bin/App1-debug.apk` to install the freshlybuilt `.apk` on your device. Start the application from your launcher (**AntApp1**) and enjoy the black screen. You can use the **BACK** key to exit the application.

Creating a Gradle-based application template manually

Gradle is a more versatile Java building tool compared to Ant, which lets you handle external dependencies and repositories with ease.

 We recommend that you watch this video from Google available at `https://www.youtube.com/watch?v=LCJAgPkpmR0` and read this official command-line building manual available at `http://developer.android.com/tools/building/building-cmdline.html` before proceeding with Gradle.

The recent versions of Android SDK are tightly integrated with Gradle, and Android Studio is built using it as its build system. Let's extend our previous `1_AntApp` application to make it buildable with Gradle.

First, go to the root folder of the project, and create the `build.gradle` file with the following content:

```
buildscript {
  repositories {
    mavenCentral()
  }
  dependencies {
    classpath 'com.android.tools.build:gradle:1.0.0'
  }
}
apply plugin: 'com.android.application'
android {
  buildToolsVersion "19.1.0"
  compileSdkVersion 19
  sourceSets {
    main {
      manifest.srcFile 'AndroidManifest.xml'
      java.srcDirs = ['src']
```

```
          resources.srcDirs = ['src']
          aidl.srcDirs = ['src']
          renderscript.srcDirs = ['src']
          res.srcDirs = ['res']
          assets.srcDirs = ['assets']
        }
      }
      lintOptions {
        abortOnError false
      }
    }
```

After this, run the command `gradle init`. The output should be similar to the following:

```
>gradle init
:init
The build file 'build.gradle' already exists. Skipping build
initialization.
:init SKIPPED
BUILD SUCCESSFUL
Total time: 5.271 secs
```

The subfolder `.gradle` will be created in the current folder. Now, run the following command:

```
>gradle build
```

The tail of the output should look as follows:

```
:packageRelease
:assembleRelease
:assemble
:compileLint
:lint
Ran lint on variant release: 1 issues found
Ran lint on variant debug: 1 issues found
Wrote HTML report to
file:/F:/Book_MasteringNDK/Sources/Chapter1/2_GradleApp/build/outputs
/lint-results.html
```

```
Wrote XML report to
F:\Book_MasteringNDK\Sources\Chapter1\2_GradleApp\build\outputs\lint-
results.xml
```

```
:check
```

```
:build
```

```
BUILD SUCCESSFUL
```

```
Total time: 9.993 secs
```

The resulting .apk packages can be found in the build\outputs\apk folder.
Try installing and running 2_GradleApp-debug.apk on your device.

Embedding native code

Let's stick to the topic of this book and write some native C++ code for our template
application. We will start with the jni/Wrappers.cpp file, which will contain a
single function definition:

```cpp
#include <stdlib.h>
#include <jni.h>
#include <android/log.h>
#define LOGI(...) ((void)__android_log_print(ANDROID_LOG_INFO,
  "NDKApp", __VA_ARGS__))
extern "C"
{
  JNIEXPORT void JNICALL
  Java_com_packtpub_ndkmastering_AppActivity_onCreateNative( JNIEnv*
    env, jobject obj )
  {
    LOGI( "Hello Android NDK!" );
  }
}
```

This function will be called from Java using the JNI mechanism. Update
AppActivity.java as follows:

```java
package com.packtpub.ndkmastering;
import android.app.Activity;
import android.os.Bundle;
public class AppActivity extends Activity
{
  static
  {
```

```
    System.loadLibrary( "NativeLib" );
  }
  @Override protected void onCreate( Bundle icicle )
  {
    super.onCreate( icicle );
    onCreateNative();
  }
  public static native void onCreateNative();
};
```

Now, we have to build this code into an installable .apk package. We need a couple of configuration files for this. The first one, jni/Application.mk, contains the platform and toolchain information:

```
APP_OPTIM := release
APP_PLATFORM := android-19
APP_STL := gnustl_static
APP_CPPFLAGS += -frtti
APP_CPPFLAGS += -fexceptions
APP_CPPFLAGS += -DANDROID
APP_ABI := armeabi-v7a-hard
APP_MODULES := NativeLib
NDK_TOOLCHAIN_VERSION := clang
```

We use the latest version of the Clang compiler—that is 3.6, as we write these lines, and the armeabi-v7a-hard target, which enables support of hardware floating point computations and function arguments passing via hardware floating point registers resulting in a faster code.

The second configuration file is jni/Android.mk, and it specifies which .cpp files we want to compile and what compiler options should be there:

```
TARGET_PLATFORM := android-19
LOCAL_PATH := $(call my-dir)
include $(CLEAR_VARS)
LOCAL_MODULE := NativeLib
LOCAL_SRC_FILES += Wrappers.cpp
LOCAL_ARM_MODE := arm
COMMON_CFLAGS := -Werror -DANDROID -DDISABLE_IMPORTGL
ifeq ($(TARGET_ARCH),x86)
  LOCAL_CFLAGS := $(COMMON_CFLAGS)
else
```

```
    LOCAL_CFLAGS := -mfpu=vfp -mfloat-abi=hard -mhard-float -fno-
      short-enums -D_NDK_MATH_NO_SOFTFP=1 $(COMMON_CFLAGS)
  endif
  LOCAL_LDLIBS := -llog -lGLESv2 -Wl,-s
  LOCAL_CPPFLAGS += -std=gnu++11
  include $(BUILD_SHARED_LIBRARY)
```

Here, we link against OpenGL ES 2, specify compiler switches to enable the hardware floating point for non-x86 targets and list the required .cpp source files.

Use the following command from the root folder of the project to build the native code:

>ndk-build

The output should be as follows:

```
>ndk-build
[armeabi-v7a-hard] Compile++ arm  : NativeLib <= Wrappers.cpp
[armeabi-v7a-hard] SharedLibrary  : libNativeLib.so
[armeabi-v7a-hard] Install        : libNativeLib.so => libs/armeabi-
v7a/libNativeLib.so
```

The last thing is to tell Gradle that we want to pack the resulting native library into the .apk. Edit the build.gradle file and add the following line to the main section of sourceSets:

jniLibs.srcDirs = ['libs']

Now, if we run the command gradle build, the resulting package build\outputs\apk\3_NDK-debug.apk will contain the required libNativeLib.so file. You can install and run it as usual. Check the line **Hello Android NDK!** printed into the Android system log with adb logcat.

> Those who do not want to tackle Gradle in such a small project without dependencies will be able to use good old Apache Ant. Just run the command ant debug to make it happen. No additional configuration files are required to put shared C++ libraries into .apk this way.

Building and signing release Android applications

We learned how to use the command line to create Android applications with the native code. Let's put the final stroke on the topic of the command-line tools and learn how to prepare and sign the release version of your application.

The detailed explanation of the signing procedure on Android is given in the developer manual at `http://developer.android.com/tools/publishing/app-signing.html`. Let's do it using Ant and Gradle.

First of all, we need to rebuild the project and create a release version of the `.apk` package. Let's do it with our 3_NDK project. We invoke `ndk-build` and Apache Ant using the following commands:

```
>ndk-build
```

```
>ant release
```

The tail of the output from Ant looks as follows:

```
-release-nosign:

[echo] No key.store and key.alias properties found in
build.properties.

[echo] Please sign
F:\Book_MasteringNDK\Sources\Chapter1\3_NDK\bin\App1-release-
unsigned.apk manually

[echo] and run zipalign from the Android SDK tools.

[propertyfile] Updating property file:
F:\Book_MasteringNDK\Sources\Chapter1\3_NDK\bin\build.prop

[propertyfile] Updating property file:
F:\Book_MasteringNDK\Sources\Chapter1\3_NDK\bin\build.prop

[propertyfile] Updating property file:
F:\Book_MasteringNDK\Sources\Chapter1\3_NDK\bin\build.prop

[propertyfile] Updating property file:
F:\Book_MasteringNDK\Sources\Chapter1\3_NDK\bin\build.prop

-release-sign:

-post-build:

release:

BUILD SUCCESSFUL

Total time: 2 seconds
```

Let's do the same thing with Gradle. Maybe you have already noticed when we run gradle build there is a 3_NDK-release-unsigned.apk file in the build/outputs/apk folder. This is exactly what we need. This will be our raw material for the signing procedure.

Now, we need to have a valid release key. We can create a self-signed release key using keytool from the Java Development Kit using the following command:

```
$ keytool -genkey -v -keystore my-release-key.keystore -alias
alias_name -keyalg RSA -keysize 2048 -validity 10000
```

This will ask us to fill out all the fields necessary for the key:

```
Enter keystore password:
Re-enter new password:
What is your first and last name?
  [Unknown]:  Sergey Kosarevsky
What is the name of your organizational unit?
  [Unknown]:  SD
What is the name of your organization?
  [Unknown]:  Linderdaum
What is the name of your City or Locality?
  [Unknown]:  St.Petersburg
What is the name of your State or Province?
  [Unknown]:  Kolpino
What is the two-letter country code for this unit?
  [Unknown]:  RU
Is CN=Sergey Kosarevsky, OU=SD, O=Linderdaum, L=St.Petersburg,
ST=Kolpino, C=RU correct?
  [no]:  yes
Generating 2048 bit RSA key pair and self-signed certificate
(SHA1withRSA) with a validity of 10000 days
for: CN=Sergey Kosarevsky, OU=SD, O=Linderdaum, L=St.Petersburg,
ST=Kolpino, C=RU
Enter key password for <alias_name>
  (RETURN if same as keystore password):
[Storing my-release-key.keystore]
```

Now, we are ready to proceed with the actual .apk package signing. Use the jarsigner tool from the Java Development Kit to do this:

```
>jarsigner -verbose -sigalg MD5withRSA -digestalg SHA1 -keystore my-
release-key.keystore 3_NDK-release-unsigned.apk alias_name
```

This command is interactive, and it will require the user to enter the keystore and the key passwords. However, we can provide both passwords as arguments to this command in the following way:

```
>jarsigner -verbose -sigalg MD5withRSA -digestalg SHA1 -keystore my-
release-key.keystore -storepass 123456 –keypass 123456 3_NDK-release-
unsigned.apk alias_name
```

Of course, passwords should match with what you have entered while creating your release key and keystore.

There is one more important thing left before we can safely proceed with publishing our .apk package on Google Play. Android applications can access uncompressed content within .apk using memory-mapped files and mmap() system calls, yet mmap() may imply some alignment restrictions on the underlying data. We need to align all uncompressed data within .apk on 4-byte boundaries. The Android SDK has the zipalign tool to do this, as seen in the following command:

```
>zipalign -v 4 3_NDK-release-unsigned.apk 3_NDK-release.apk
```

Now, our .apk is ready to be published on Google Play.

Organizing the cross-platform code

This book continues the idea from our previous book *Android NDK Game Development Cookbook, Packt Publishing*: the possibility of cross-platform development using the principle *What You See (on a desktop PC) is What You Get (on a mobile device)*. Most of the application logic can be developed and tested in a familiar desktop environment such as Windows with all necessary tools at hand, and this can be built for Android using the NDK whenever necessary.

To organize and maintain the cross-platform C++ source code, we need to split everything into platform-specific and platform-independent parts. Our Android-specific native code will be stored in the jni subfolder of the project, exactly as we did in our previous minimalistic example. The shared platform-independent C++ code will go into the src-native subfolder.

Using TeamCity continuous integration server with Android applications

TeamCity is a powerful continuous integration and deployment server, which can be used to automate your Android application builds. This can be found at https://www.jetbrains.com/teamcity.

 TeamCity is free for small projects that require no more than 20 build configurations and 3 build agents and is absolutely free for open source projects. Ask for an open-source license at https://www.jetbrains.com/teamcity/buy.

The server installation procedure is straightforward. Windows, OS X, or Linux machine can be used as the server or a build agent. Here, we will show how to install TeamCity on Windows.

Download the latest version of the installer from https://www.jetbrains.com/teamcity/download and run it using the following command:

```
>TeamCity-9.0.1.exe
```

Install all components and run it as **Windows Service**. For simplicity, we will run both the server and the agent on a single machine, as shown in the following screenshot:

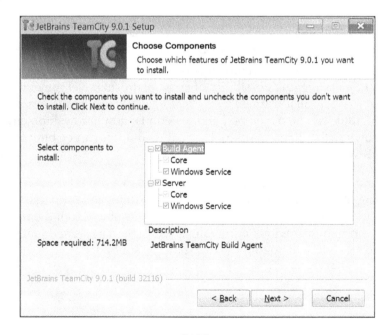

Choose the desired TeamCity server port. We will use the default HTTP port 80. Run the **TeamCity Server** and **Agent** services under the SYSTEM account.

Once the server is online, open your browser and connect to it using the address `http://localhost`. Create a new project and a build configuration.

 To work with TeamCity, you should put the source code of your project into a version control system. Git and GitHub will be a good choice.

If your project is already on GitHub, you can create a Git VCS root pointing to the URL of your GitHub repository, like this `https://github.com/<your login>/<your project>.git`.

Add a new command-line build step and type the content of the script:

```
ndk-build
ant release
```

You can also add signing using `jarsigner` here and use the `zipalign` tool to create the final production `.apk`.

Now, go to the **General Settings** step and add an artifact path to `bin/3_NDK-release.apk`. The project is ready for continuous integration.

Summary

In this chapter, we learned how to install and configure the essential tools for Android native development using the command line, and how to write Android application basic configuration files manually without falling back to graphical IDEs. In the subsequent chapters, we will practice these skills and build some projects.

2
Native Libraries

In this chapter, you will learn how to build popular C/C++ libraries and link them against your applications using Android NDK. These libraries are building blocks to implement feature-rich applications with images, videos, sounds, physical simulations, and networking entirely in C++. We will provide minimal samples to demonstrate the functionality of each library. Audio and networking libraries are discussed in greater detail in the subsequent chapters. We will show you how to compile libraries and, of course, give some short samples and hints on how to start using them.

Typical caveats for porting libraries across different processors and operating systems are memory access (structure alignment and padding), byte order (endianness), calling conventions, and floating-point issues. All the libraries described in the preceding sections address these issues quite well, and even if some of them do not officially support Android NDK, fixing this is just a matter of a few compiler switches.

To build the mentioned libraries, we need to create makefiles for Windows, Linux, and OS X and a pair of the Android.mk/Application.mk files for the NDK. The source files of the library are compiled to object files. A collection of object files is combined into an archive, which is also called a static library. Later, this static library can be passed as an input to the linker. We start with the desktop versions, first for Windows.

To build the Windows-specific version of libraries, we need a C++ compiler. We will use the command line compiler from MinGW with the GCC toolchain described in *Chapter 1, Using Command-line Tools*. For each library, we have a collection of source code files, and we need to get the static library, a file with the .a extension.

Dealing with precompiled static libraries

Let's put the source code of a library we need to build for different platforms into the src directory. The Makefile script should start as follows:

```
CFLAGS = -O2 -I src
```

This line defines a variable CFLAGS with a list of compiler command-line parameters. In our case, we instruct the compiler to search the src directory for header files. If the library source code spans across multiple directories, we need to add the -I switch for each of the directories. The -O2 switch tells the compiler to enable level 2 optimizations. Next, we add the following lines for each source file:

```
<SourceFileName>.o:
gcc $(CFLAGS) -c <SourceFile>.cpp -o <SourceFile>.o
```

The string <SourceFileName> should be replaced with the actual name of the .cpp source file, and these lines should be written for each of the source files.

Now, we add the list of object files:

```
ObjectFiles = <SourceFile1>.o <SourceFile2>.o
```

Finally, we will write the target for our library:

```
<LibraryName>:
ar -rvs <LibraryName>.a $(ObjectList)
```

Every line in the Makefile script except the empty lines and the names of the targets should start with a tabulation character. To build the library, invoke the following command:

```
>make <LibraryName>.a
```

When using the library in our programs, we pass the LibraryName.a file as a parameter to gcc.

Makefiles consist of targets similar to subroutines in programming languages, and usually each target results in an object file being generated. For example, we have seen that each source file of the library gets compiled into the corresponding object file.

Target names may include the filename pattern to avoid copying and pasting, but in the simplest case, we just list all the source files and duplicate those lines replacing the `SourceFileName` strings by the appropriate file names. The `-c` switch after the `gcc` command is the option to compile the source file, and `-o` specifies the name of the output object file. The `$(CFLAGS)` symbol denotes the substitution of the value of the `CFLAGS` variable into the command line.

The GCC toolchain for Windows includes the `ar` tool, which is an abbreviation for an archiver. Makefiles for our libraries invoke this tool to create a static version of the library. This is done in the last lines of the Makefile script.

When a line with a list of object files becomes too long, it can be split into multiple lines using the backslash symbol as follows:

```
ObjectFileList = FileName1.o \
           ... \
           FileNameN.o
```

There should be no spaces after the backslash, since it is a limitation of the `make` tool. The `make` tool is portable, hence the same rules apply exactly to all desktop operating systems we use: Windows, Linux, and OS X.

Now, we are able to build most of the libraries using Makefiles and the command line. Let's build them for Android. First, create a folder named `jni` and create the `jni/Application.mk` file with the appropriate compiler switches and set the name of the library accordingly. For example, one for the Theora library should look like the following:

```
APP_OPTIM := release
APP_PLATFORM := android-19
APP_STL := gnustl_static
APP_CPPFLAGS += -frtti
APP_CPPFLAGS += -fexceptions
APP_CPPFLAGS += -DANDROID
APP_ABI := armeabi-v7a-hard
APP_MODULES := Theora
NDK_TOOLCHAIN_VERSION := clang
```

Here, we will use `armeabi-v7a-hard` as one of the most widely supported modern ABIs. The Android NDK supports many other architectures and CPUs. Refer to the NDK Programmer's Guide for a complete up-to-date list.

It will use the latest version of the Clang compiler available in the installed NDK. The `jni/Android.mk` file is similar to the one we wrote in the previous chapter for the `3_NDK` sample application, yet with a few exceptions. At the top of the file, some required variables must be defined. Let's see how the `Android.mk` file for the OpenAL-Soft library might look:

```
TARGET_PLATFORM := android-19
LOCAL_PATH := $(call my-dir)
include $(CLEAR_VARS)
LOCAL_ARM_MODE := arm
LOCAL_MODULE := OpenAL
LOCAL_C_INCLUDES += src
LOCAL_SRC_FILES += <ListOfSourceFiles>
```

Define some common compiler options: treat all warnings as errors (`-Werror`), the `ANDROID` preprocessing symbol is defined:

```
COMMON_CFLAGS := -Werror -DANDROID
```

The compilation flags are defined according to the selected CPU architecture:

```
ifeq ($(TARGET_ARCH),x86)
  LOCAL_CFLAGS := $(COMMON_CFLAGS)
else
  LOCAL_CFLAGS := -mfpu=vfp -mfloat-abi=hard -mhard-float -fno-
    short-enums -D_NDK_MATH_NO_SOFTFP=1 $(COMMON_CFLAGS)
endif
```

In all our examples, we will use the hardware floating point ABI `armeabi-v7a-hard`, so let's build the libraries accordingly.

 The major difference between armeabi-v7a-hard and armeabi-v7a is that the hardware floating point ABI does passing of the floating point function arguments inside FPU registers. In floating point heavy applications, this can drastically increase the performance of the code where floating point values are passed to and from different functions.

Since we are building a static library, we need the following line at the end of `Android.mk`:

```
include $(BUILD_STATIC_LIBRARY)
```

Building the static library now requires a single call to the `ndk-build` script. Let's proceed to the compilation of actual libraries after a small remark on dynamic linking and Windows platform.

Dynamic linking on Windows platform

The libraries considered in this chapter can be built for Windows as dynamic link libraries. We do not provide recipes for doing this because each project already contains all the necessary instructions, and Windows development is not the focus of this book. The only exception is the libcurl and OpenSSL libraries. We recommend that you download the prebuilt DLL files from the official library site.

In the example code for FreeImage, FreeType, and Theora, we use function pointers, which are initialized using the GetProcAddress() and LoadLibrary() functions from WinAPI. The same function pointers are used on Android, but in this case, they point to appropriate functions from a static library.

For example, the function FreeImage_OpenMemory() is declared as follows:

```
typedef FIMEMORY* ( DLL_CALLCONV* PFNFreeImage_OpenMemory )
  ( void*, unsigned int );
PFNFreeImage_OpenMemory  FI_OpenMemory = nullptr;
```

On Windows, we initialize the pointer with the GetProcAddress() call:

```
FI_OpenMemory = (PFNFreeImage_OpenMemory)
  GetProcAddress (hFreeImageDLL, "FreeImage_OpenMemory");
```

On Android, OSX, and Linux, it is a redirection:

```
FI_OpenMemory = &FreeImage_OpenMemory;
```

The example code only refers to FI_OpenMemory(), and thus, is the same for both Android and Windows.

Curl

The libcurl library http://curl.haxx.se/libcurl is a free and easy to use client-side URL transfer library. It is a de facto standard for native applications, which deal with numerous networking protocols. Linux and OS X users enjoy having the library on their system, and a possibility to link against it using the -lcurl switch. Compilation of libcurl for Android on a Windows host requires some additional steps to be done. We explain them here.

The libcurl library build process is based on autoconf; we will need to generate the curl_config.h file before actually building the library. Run the configure script from the folder containing the unpacked libcurl distribution package. Cross-compilation command-line flags should be set to:

```
--host=arm-linux CC=arm-eabi-gcc
```

The `-I` parameter of the `CPPFLAGS` variable should point to the `/system/core/include` subfolder of your NDK folder, in our case:

```
CPPFLAGS="-I D:/NDK/system/core/include"
```

The libcurl library can be customized in many ways. We use this set of parameters (disable all protocols except HTTP and HTTPS):

```
>configure CC=arm-eabi-gcc --host=arm-linux --disable-tftp --disable-
sspi --disable-ipv6 --disable-ldaps --disable-ldap --disable-telnet -
-disable-pop3 --disable-ftp --without-ssl --disable-imap --disable-
smtp --disable-pop3 --disable-rtsp --disable-ares --without-ca-bundle
--disable-warnings --disable-manual --without-nss --enable-shared --
without-zlib --without-random --enable-threaded-resolver --with-ssl
```

The `--with-ssl` parameter enables the usage of OpenSSL library to provide secure HTTPS transfers. This library will be discussed further in this chapter. However, in order to work with SSL-encrypted connections, we need to tell libcurl where our system certificates are located. This can be done with `CURL_CA_BUNDLE` defined in the beginning of the `curl_config.h` file:

```
#define CURL_CA_BUNDLE "/etc/ssl/certs/ca-certificates.crt"
```

The configure script will generate a valid `curl_config.h` header file. You may find it in the book's source code bundle. Compilation of an Android static library requires a usual set of `Android.mk` and `Application.mk` files, which is also within the `1_Curl` example. In the next chapter, we will learn how to use the libcurl library to download the actual content from Internet over HTTPS. However, here is a simplistic usage example to retrieve a HTTP page:

```
CURL* Curl = curl_easy_init();
curl_easy_setopt( Curl, CURLOPT_URL, "http://www.google.com" );
curl_easy_setopt( Curl, CURLOPT_FOLLOWLOCATION, 1 );
curl_easy_setopt( Curl, CURLOPT_FAILONERROR, true );
curl_easy_setopt( Curl, CURLOPT_WRITEFUNCTION, &MemoryCallback );
curl_easy_setopt( Curl, CURLOPT_WRITEDATA, 0 );
curl_easy_perform( Curl );
curl_easy_cleanup( Curl );
```

Here `MemoryCallback()` is a function that handles the received data. It can be as tiny as the following code fragment:

```
size_t MemoryCallback( void* P, size_t Size, size_t Num, void* )
{
  if ( !P ) return 0;
  printf( "%s\n", P );
}
```

The retrieved data will be printed on the screen in your desktop application. The same code will work like a dummy in Android, without producing any visible side effects, since the `printf()` function is just a dummy there.

OpenSSL

OpenSSL is an open source library implementing the Secure Sockets Layer (SSL v2/v3) and Transport Layer Security (TLS) protocols as well as a full-strength general purpose cryptography library. It can be found at `https://www.openssl.org`.

Here, we will build the OpenSSL Version 1.0.1j, which contains a fix for the Heartbleed Bug(`http://heartbleed.com`).

The Heartbleed Bug is a serious vulnerability in the popular OpenSSL cryptographic software library. This weakness allows stealing the information that is protected, under normal conditions, by the SSL/TLS encryption used to secure the Internet.

If you try to statically link your application against an old version of OpenSSL and then publish it on Google Play, you may see the following security alert:

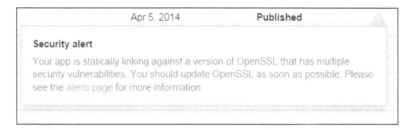

It is possible that by the time this book is published, even the version 1.0.0j of OpenSSL will be outdated. Hence, it would be a great exercise for you to download the most recent source code and update NDK Makefiles accordingly. Here is a brief glimpse of how you can do it.

OpenSSL is compiled as two interoperating static libraries: `libssl` and `libcrypto`. Check out the source code bundle and look into the folders `2_OpenSSL/lib/crypto/jni` and `2_OpenSSL/ssl/jni`. Both libraries should be linked against your application which uses SSL-enabled version of libcurl.

Typical `Android.mk` for this can start as in the following listing:

```
include $(CLEAR_VARS)
LOCAL_MODULE := libCurl
LOCAL_SRC_FILES :=
  ../../../Libs.Android/libcurl.$(TARGET_ARCH_ABI).a
include $(PREBUILT_STATIC_LIBRARY)
include $(CLEAR_VARS)
LOCAL_MODULE := libCrypto
LOCAL_SRC_FILES :=
  ../../../Libs.Android/libCrypto.$(TARGET_ARCH_ABI).a
include $(PREBUILT_STATIC_LIBRARY)
include $(CLEAR_VARS)
LOCAL_MODULE := libSSL
LOCAL_SRC_FILES :=
  ../../../Libs.Android/libSSL.$(TARGET_ARCH_ABI).a
include $(PREBUILT_STATIC_LIBRARY)
```

At the end of this file, just link all the libraries:

```
LOCAL_STATIC_LIBRARIES += libCurl
LOCAL_STATIC_LIBRARIES += libSSL
LOCAL_STATIC_LIBRARIES += libCrypto
```

This is it, you can now deal with SSL connections.

FreeImage

FreeImage is a popular library for bitmap manipulation, Unity gaming engine is among the users of this library (`http://freeimage.sourceforge.net/users.html`). The library is an all-in-one wrapper on top of `libpng`, `libjpeg`, `libtiff`, and many others, providing fast image loading routines without falling back to Java code.

FreeImage includes a complete set of Makefiles for different platforms. The compilation of the library for Android is straightforward with the instructions from *Dealing with precompiled static libraries* section. The `Application.mk` file differs from the same file for Curl in one line:

```
APP_MODULES := FreeImage
```

In the `Android.mk` file, we will change the C compilation flags:

```
GLOBAL_CFLAGS   := -O3 -DHAVE_CONFIG_H=1 -DFREEIMAGE_LIB -
  DDISABLE_PERF_MEASUREMENT
```

In the following sample, we will implement two simple routines to load and save images in various file formats to and from memory blocks.

We start with the `FreeImage_LoadFromMemory()` routine, which takes the `Data` array and its `Size` as input parameters and decodes this array into a `std::vector<char>` containing bitmap's pixels. Dimensions information, width and height, is stored in the `W` and `H` parameters. Color depth information is put into the `BitsPerPixel` parameter. An optional `DoFlipV` parameter instructs the code to flip the loaded image vertically, this can be required when dealing with images storing conventions in different graphics APIs, top-down or bottom-up:

```
bool FreeImage_LoadFromStream( void* Data,unsigned int Size,
  std::vector<ubyte>& OutData,int& W,
  int& H,int& BitsPerPixel,bool DoFlipV )
{
```

We create the internal memory block, which can be read by FreeImage routines:

```
FIMEMORY* Mem = FI_OpenMemory(
  ( unsigned char* )Data,
  static_cast<unsigned int>( Size )
);
```

Before reading the bitmap, we will detect its format (for example, `.jpg`, `.bmp`, `.png`, and others) in the following way:

```
FREE_IMAGE_FORMAT FIF = FI_GetFileTypeFromMemory( Mem, 0 );
```

Then, the decoded bitmap is read into the temporary `FIBITMAP` structure:

```
FIBITMAP* Bitmap = FI_LoadFromMemory( FIF, Mem, 0 );
FI_CloseMemory( Mem );
FIBITMAP* ConvBitmap;
```

If the total number of bits is more than 32, for example, each color channel occupies more than 8 bits, we most likely have the floating-point image, which will require some additional processing:

```
bool FloatFormat = FI_GetBPP( Bitmap ) > 32;
if ( FloatFormat )
{
```

Floating point images are not used throughout this book, but it is useful to know FreeImage supports high dynamic range images in the OpenEXR format.

 OpenEXR format is notable for supporting 16-bit-per-channel floating point values and can be used in games to store textures for different HDR effect.

```
ConvBitmap = FI_ConvertToRGBF( Bitmap );
}
else
{
```

A transparency information is used to convert the image. If the image is not transparent, the alpha channel is ignored:

```
ConvBitmap = FI_IsTransparent( Bitmap ) ? FI_ConvertTo32Bits(
   Bitmap ) : FI_ConvertTo24Bits( Bitmap );
}
FI_Unload( Bitmap );
Bitmap = ConvBitmap;
```

If necessary, we do the vertical flipping of the image, as follows:

```
if ( DoFlipV ) FI_FlipVertical( Bitmap );
```

The image dimensions and color information are extracted:

```
W = FI_GetWidth( Bitmap );
H = FI_GetHeight( Bitmap );
BitsPP = FI_GetBPP( Bitmap );
```

Once we know the dimensions, we can resize the output buffer, as follows:

```
OutData.resize( W * H * ( BitsPerPixel / 8 ) );
```

At last, we can fetch the raw unaligned bitmap data to our OutData vector. The size of a single tightly packed scanline is W*BitsPP/8 bytes:

```
FI_ConvertToRawBits( &OutData[0],Bitmap, W * BitsPP / 8, BitsPP,
   0, 1, 2, false );
```

The temporary bitmap object is deleted and function returns gracefully:

```
FI_Unload( Bitmap );
return true;
}
```

The bitmap saving routine can be implemented in a similar way. First, we allocate the FIBITMAP structure to represent our image within the FreeImage library:

```
bool FreeImage_SaveToMemory( const std::string& Ext,
  ubyte* RawBGRImage,int Width,int Height,int BitsPP,
  std::vector<ubyte>& OutData )
{
  FIBITMAP* Bitmap = FI_Allocate(Width, Height, BitsPP, 0, 0, 0);
```

Raw bitmap data is copied into the FIBITMAP structure:

```
memcpy( FI_GetBits( Bitmap ),
  RawBGRImage, Width * Height * BitsPP / 8 );
```

FreeImage uses the inverted vertical scanline order, so we should flip the image vertically before saving:

```
FI_FlipVertical( Bitmap );
```

Then, we will use the user-specified file extension to detect the format of the output image:

```
int OutSubFormat;
FREE_IMAGE_FORMAT OutFormat;
FileExtToFreeImageFormats( Ext, OutSubFormat, OutFormat );
```

To save the image, we will allocate a dynamic memory block:

```
FIMEMORY* Mem = FI_OpenMemory( nullptr, 0);
```

The FI_SaveToMemory() call encodes our raw bitmap into the compressed representation according to the selected format:

```
if ( !FI_SaveToMemory( OutFormat,
  Bitmap, Mem, OutSubFormat ) )
{
  return false;
}
```

After encoding, we will get direct access to the FreeImage memory block:

```
ubyte* Data = NULL;
uint32_t Size = 0;
FI_AcquireMemory( Mem, &Data, &Size );
```

Then, we will copy bytes to our OutData vector:

```
OutData.resize( Size );
memcpy( &OutData[0], Data, Size );
```

Some cleanup is required. We delete the memory block and the FIBITMAP structure:

```
FI_CloseMemory( Mem );
FI_Unload( Bitmap );
return true;
}
```

The auxiliary `FileExtToFreeImageFormats()` function converts the file extension to the internal FreeImage format specifier and provides a number of options. The code is straightforward. We will compare the provided file extension to a number of predefined values and fill the FIF_FORMAT and SAVE_OPTIONS structures:

```
static void FileExtToFreeImageFormats( std::string Ext,
  int& OutSubFormat, FREE_IMAGE_FORMAT& OutFormat )
{
  OutSubFormat = TIFF_LZW;
  OutFormat = FIF_TIFF;
  std::for_each( Ext.begin(), Ext.end(),
    []( char& in )
  {
    in = ::toupper( in );
  }
  );
  if ( Ext == ".PNG" )
  {
    OutFormat = FIF_PNG;
    OutSubFormat = PNG_DEFAULT;
  }
  else if ( Ext == ".BMP" )
  {
    OutFormat = FIF_BMP;
    OutSubFormat = BMP_DEFAULT;
  }
  else if ( Ext == ".JPG" )
  {
    OutFormat = FIF_JPEG;
    OutSubFormat = JPEG_QUALITYSUPERB | JPEG_BASELINE |
      JPEG_PROGRESSIVE | JPEG_OPTIMIZE;
  }
  else if ( Ext == ".EXR" )
  {
    OutFormat = FIF_EXR;
    OutSubFormat = EXR_FLOAT;
  }
}
```

This can be extended and customized further in your own way.

Loading and saving images

To make the preceding code usable, we add two more routines that save and load images from disk files. The first one, `FreeImage_LoadBitmapFromFile()`, loads the bitmap:

```
bool FreeImage_LoadBitmapFromFile( const std::string& FileName,
  std::vector<ubyte>& OutData, int& W, int& H, int& BitsPP )
{
  std::ifstream InFile( FileName.c_str(),
  std::ios::in | std::ifstream::binary );
  std::vector<char> Data(
    ( std::istreambuf_iterator<char>( InFile ) ),
    std::istreambuf_iterator<char>() );
  return FreeImage_LoadFromStream(
    ( ubyte* )&Data[0], ( int )data.size(),
    OutData, W, H, BitsPP, true );
}
```

We use a simple function to extract the file extension, which serves as a file type tag:

```
std::string ExtractExtension( const std::string& FileName )
{
  size_t pos = FileName.find_last_of( '.' );
  return ( pos == std::string::npos ) ?
    FileName : FileName.substr( pos );
}
```

The `FreeImage_SaveBitmapToFile()` function saves the file using the standard `std::ofstream` stream:

```
bool FreeImage_SaveBitmapToFile( const std::string& FileName,
  ubyte* ImageData, int W, int H, int BitsPP )
{
  std::string Ext = ExtractExtension( FileName );
  std::vector<ubyte> OutData;
  if ( !FreeImage_SaveToMemory( Ext, ImageData, W, H, BitsPP,
    OutData ) )
  {
    return false;
  }
  std::ofstream OutFile( FileName.c_str(),
  std::ios::out | std::ofstream::binary );
  std::copy( OutData.begin(), OutData.end(),
    std::ostreambuf_iterator<char>( OutFile ) );
  return true;
}
```

This code is enough to cover all the basic use cases of the image loading library.

FreeType

The FreeType library is a de facto standard to render high-quality text using TrueType fonts. Since text output is almost inevitable in any graphical program, we give an example how to render a text string using a fixed-size font generated from the monospace TrueType file.

We store the fixed-size font in the `16x16` grid. The source font for this demo application is named `Receptional Receipt` and was downloaded from `http://1001freefonts.com`. Four lines of the resulting `16x16` grid are shown in the following image:

A single character occupies a rectangular region, which we will call a *slot*. The coordinates of the character's rectangle are calculated using the character's ASCII code. Each slot in a grid occupies the `SlotW` x `SlotH` pixels, and the character itself is centered and has the size of `CharW` x `CharH` pixels. For demonstration purposes, we simply assume `SlotW` is two times the size of `CharW`:

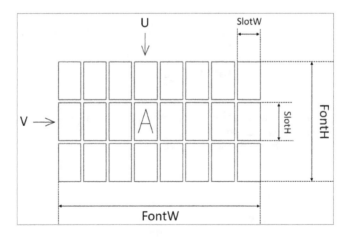

We limit ourselves to the simplest possible usage scenario: 8-bit ASCII characters, fixed-size character glyphs. To render the string, we will iterate its characters and call the yet-to-be-written function `RenderChar()`:

```
void RenderStr( const std::string& Str, int x, int y )
{
  for ( auto c: Str )
  {
    RenderChar( c, x, y );
    x += CharW;
  }
}
```

The character rendering routine is a simple double loop copying the glyph pixels into the output image:

```
void RenderChar( char c, int x, int y )
{
  int u = ( c % 16 ) * SlotW;
  int v = ( c / 16 ) * SlotH;
  for ( int y1 = 0 ; y1 < CharH ; y1++ )
    for ( int x1 = 0 ; x1 <= CharW ; x1++ )
      PutPixel( g_OutBitmap, W, H,
        x + x1, y + y1,
        GetPixel( Font, FontW, FontH,
          x1 + u + CharW, y1 + v)
      );
}
```

The `PutPixel()` and `GetPixel()` routines set and get the pixel in the bitmap respectively. Each pixel is in the 24-bit RGB format:

```
int GetPixel( const std::vector<unsigned char>& Bitmap,
  int W, int H, int x, int y )
{
  if ( y >= H || x >= W || y < 0 || x < 0 ) { return 0; }
```

Here, a scanline width is assumed to be equal to the image width, and the number of color components in the RGB triplet is 3:

```
int Ofs = ( y * W + x ) * 3;
```

Use bitwise shifts to construct the resulting RGB value:

```
return (Bitmap[Ofs+0] << 16) +
  (Bitmap[Ofs+1] <<  8) +
  (Bitmap[Ofs+2]);
}

void PutPixel( std::vector<unsigned char>& Bitmap,
  int W, int H, int x, int y, int Color )
{
  if ( y < 0 || x < 0 || y > H - 1 || x > W - 1 ) { return; }
  int Ofs = ( y * W + x ) * 3;
```

Bitwise shifts and masks do the extraction job:

```
buffer[Ofs + 0] = ( Color ) & 0xFF;
buffer[Ofs + 1] = ( Color >> 8 ) & 0xFF;
buffer[Ofs + 2] = ( Color >> 16 ) & 0xFF;
}
```

There is another auxiliary function Greyscale(), which calculates the RGB gray color for a given intensity using bitwise shifts:

```
inline int Greyscale( unsigned char c )
{
  return ( (255-c) << 16 ) + ( (255-c) << 8 ) + (255-c);
}
```

For the preceding code, we do not require FreeType. We really need the library only to generate the font. We will load the font data file, render its glyphs for the first 256 characters, and then use the resulting font bitmap to render the text string. The first part of the code generates a font. We will use a few of the variables to store the dimensions of the font:

```
/// Horizontal size of the character
const int CharW = 32;
const int CharH = 64;
/// Horizontal size of the character slot
const int SlotW = CharW * 2;
const int SlotH = CharH;
const int FontW = 16 * SlotW;
const int FontH = 16 * SlotH;
std::vector<unsigned char> g_FontBitmap;
```

We store the font in a standard vector, which we can pass to the
`TestFontRendering()` routine:

```
void TestFontRendering( const std::vector<char>& Data )
{
  LoadFreeImage();
  LoadFreeType();
  FT_Library Library;
  FT_Init_FreeTypePTR( &Library );
  FT_Face Face;
  FT_New_Memory_FacePTR( Library,
    (const FT_Byte*)Data.data(),
    (int)Data.size(), 0, &face );
```

Fix the character size at 100 dpi:

```
FT_Set_Char_SizePTR( Face, CharW * 64, 0, 100, 0 );
g_FontBitmap.resize( FontW * FontH * 3 );
std::fill( std::begin(g_FontBitmap),
  std::end(g_FontBitmap), 0xFF );
```

We will render 256 ASCII characters one by one in a loop:

```
for ( int n = 0; n < 256; n++ )
{
```

Load the glyph image into the slot:

```
if ( FT_Load_CharPTR( Face, n , FT_LOAD_RENDER ) )
  continue;
FT_GlyphSlot Slot = Face->glyph;
FT_Bitmap Bitmap = Slot->bitmap;
```

The coordinates of the top left corner of the rectangle for each character are
calculated:

```
int x = (n % 16) * SlotW + CharW + Slot->bitmap_left;
int y = (n / 16) * SlotH - Slot->bitmap_top + 3*CharH/4;
```

The glyph of the character is copied to the `g_FontBitmap` bitmap:

```
for ( int i = 0 ; i < ( int )Bitmap.width; i++ )
for ( int j = 0 ; j < ( int )Bitmap.rows; j++ )
PutPixel( g_FontBitmap, FontW, FontH,i + x, j + y,
  Greyscale( Bitmap.buffer[j * Bitmap.width + i])
);
}
```

We save the resulting `Font` bitmap to a file:

```
FreeImage_SaveBitmapToFile( "test_font.png",
   g_FontBitmap.data(), FontW, FontH, 24 );
```

At the end of font bitmap generation, we will clear everything associated with the FreeType library:

```
FT_Done_FacePTR     ( Face );
FT_Done_FreeTypePTR( Library );
```

To utilize our monospace font, we will declare the string, calculate its width in screen pixels, and allocate the output bitmap:

```
std::string Str = "Test string";
W = Str.length() * CharW;
H = CharH;
g_OutBitmap.resize( W * H * 3 );
std::fill( std::begin(g_OutBitmap),
std::end(g_OutBitmap), 0xFF );
```

The end of the `TestFontRendering()` routine just calls `RenderStr()`:

```
RenderStr( Str, 0, 0 );
```

It then saves the resulting image to a file:

```
FreeImage_SaveBitmapToFile( "test_str.png",
   g_OutBitmap.data(), W, H, 24 );
}
```

The result should look as in the following image:

Usually when it comes to bitmap font rendering, you don't want to write the code for bitmap generation yourself. It is advised that you use third-party tools to do it. One of such free tools is AngelCode, which can be found at `http://www.angelcode.com/products/bmfont`. It can pack the glyphs inside the bitmap in an optimal way and produce the required data to handle the generated bitmap properly.

Theora

Theora is a free and open video compression format from the Xiph.Org Foundation. Like all our multimedia technology, it can be used to distribute movies and video online and on-disc without the licensing and royalty fees or any other vendor lock-ins associated with many other video formats. It is available at `http://www.theora.org`.

To avoid confusion, we will introduce some nomenclature. By **bitstream**, we assume some sequence of bytes. Logical bitstream is some representation of video or audio data. **Codec**, or COder-DECoder, is a set of functions that encodes and decodes logical bitstreams into a set of compact representations named packed bitstreams. Since usual multimedia data consists of multiple logical bitstreams, the compact representation must be split into small chunks, which are called packets. Each **packet** has a specific size, a timestamp and a checksum associated with it to guarantee the packet integrity. The scheme for bitstreams and packets is shown in the following image:

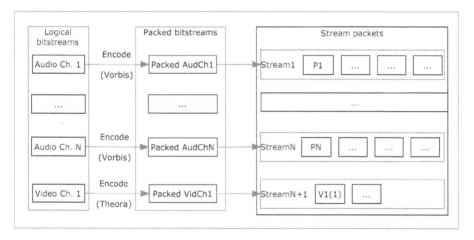

Packets for logical and packed bitstreams are intermixed to form a linear sequence maintaining the order of packets for each individual bitstream. This is called multiplexing. The Ogg library reads the `.ogg` file and splits it into packed bitstreams. Each of bitstreams can be decoded using Theora, Vorbis, or other decoders.

 In our previous book, *Android NDK Game Development Cookbook*, *Packt Publishing* (`https://www.packtpub.com/game-development/android-ndk-game-development-cookbook`), we taught by example how to decode Ogg Vorbis audio streams.

In this chapter, we address only the simplest problem of extracting media information from the file. The code for even this simple operation may seem long and complicated. However, it can be described in less than ten steps:

1. Initialize the OGG stream reader.
2. Start a packet building loop: read a bunch of bytes from the source file.
3. Check if there is enough data to emit another logical packet.
4. If the new packet is formed, check whether it is a BoS (Beginning of Stream) packet.
5. Try to initialize the Theora or Vorbis decoder with the BoS packet.
6. If we don't have enough audio and video streams to decode, go to step 2.
7. If we don't have enough stream information, continue reading secondary stream packets.
8. Initialize the Theora decoder and extract video frame information.

 There is another level of complexity in Ogg streams, because packets are grouped to form logical pages. In the preceding pseudocode, we refer to packets that are actually pages. Nevertheless, the scheme remains the same: read bytes until there is enough data for the decoder to emit another video frame or, in our case, to read the video information.

We use standard C++ I/O streams and implement three simple functions: Stream_Read(), Stream_Seek(), and Stream_Size(). Later in the *Chapter 4, Organizing a Virtual Filesystem*, we will reimplement these methods using our own I/O abstraction layer. Let's open the file stream:

```
std::ifstream Input( "test.ogv", std::ios::binary );
```

Here is a function to read the specified number of bytes from the input stream:

```
int Stream_Read( char* OutBuffer, int Size )
{
  Input.read( OutBuffer, Size );
  return Input.gcount();
}
```

Seek to the desired position using the following code:

```
int Stream_Seek( int Offset )
{
  Input.seekg( Offset );
  return (int)Input.tellg();
}
```

To determine the file size use the following code:

```
int Stream_Size()
{
  Input.seekg (0, input.end);
  int Length = Input.tellg();
  Input.seekg( 0, Input.beg );
  return Length;
}
```

At the beginning, some variables should be declared to store the state of the decoding process, a synchronization object, the current page, and audio and video streams:

```
ogg_sync_state    OggSyncState;
ogg_page          OggPage;
ogg_stream_state  VorbisStreamState;
ogg_stream_state  TheoraStreamState;
```

The Theora decoder state :

```
th_info           TheoraInfo;
th_comment        TheoraComment;
th_setup_info*    TheoraSetup;
th_dec_ctx*       TheoraDecoder;
```

The Vorbis decoder state:

```
vorbis_info       VorbisInfo;
vorbis_dsp_state  VorbisDSPState;
vorbis_comment    VorbisComment;
vorbis_block      VorbisBlock;
```

The function `Theora_Load()` reads the file header and extracts video frame information from it:

```
bool Theora_Load()
{
  Stream_Seek( 0 );
```

The current Ogg packet will be read into the structure `TempOggPacket`:

```
ogg_packet TempOggPacket;
```

Some simple yet necessary initialization of the state variables has to be done:

```
memset ( &VorbisStreamState, 0, sizeof ( ogg_stream_state ) );
memset ( &TheoraStreamState, 0, sizeof ( ogg_stream_state ) );
memset ( &OggSyncState,      0, sizeof ( ogg_sync_state ) );
memset ( &OggPage,           0, sizeof ( ogg_page ) );
memset ( &TheoraInfo,        0, sizeof ( th_info ) );
memset ( &TheoraComment,     0, sizeof ( th_comment ) );
memset ( &VorbisInfo,        0, sizeof ( vorbis_info ) );
memset ( &VorbisDSPState,    0, sizeof ( vorbis_dsp_state ) );
memset ( &VorbisBlock,       0, sizeof ( vorbis_block ) );
memset ( &VorbisComment,     0, sizeof ( vorbis_comment ) );
OGG_sync_init    ( &OggSyncState );
TH_comment_init ( &TheoraComment );
TH_info_init     ( &TheoraInfo );
VORBIS_info_init ( &VorbisInfo );
VORBIS_comment_init ( &VorbisComment );
```

We start reading the file and use the Done flag to terminate if the file has ended or we have enough data to get the information:

```
bool Done = false;
while ( !Done )
{
  char* Buffer = OGG_sync_buffer( &OggSyncState, 4096 );
  int BytesRead = ( int )Stream_Read( Buffer, 4096 );
  OGG_sync_wrote( &OggSyncState, BytesRead );
  if ( BytesRead == 0 )
  {
    break;
  }
  while (OGG_sync_pageout( &OggSyncState, &OggPage ) > 0)
  {
```

When we finally encounter a complete packet, we will check whether it is a BOS marker, and output the data to one of the decoders:

```
ogg_stream_state OggStateTest;
if ( !OGG_page_bos( &OggPage ) )
{
  if ( NumTheoraStreams > 0 )
  {
    OGG_stream_pagein( &TheoraStreamState, &OggPage );
  }
```

```
if ( NumVorbisStreams > 0 )
  {
    OGG_stream_pagein( VorbisStreamState, &OggPage );
  }
  Done = true;
  break;
}
OGG_stream_init( &OggStateTest,
OGG_page_serialno( &OggPage ) );
OGG_stream_pagein( &OggStateTest, &OggPage );
OGG_stream_packetout( &OggStateTest, &TempOggPacket );
```

We will use two variables, `NumTheoraStreams` and `NumVorbisStreams`, to count video and audio streams, respectively. In the following lines, we will feed the Ogg packet to both decoders and see if the decoders do not complain about it:

```
if ( NumTheoraStreams == 0 )
{
  int Ret = TH_decode_headerin( &TheoraInfo,
    &TheoraComment, &TheoraSetup, &TempOggPacket );
  if ( Ret > 0 )
  {
```

Here goes the Theora header:

```
    memcpy( &TheoraStreamState, &OggStateTest,
      sizeof( OggStateTest ) );
    NumTheoraStreams = 1;
    continue;
  }
}
if ( NumVorbisStreams == 0 )
{
  int Ret = VORBIS_synthesis_headerin( &VorbisInfo,
    &VorbisComment, &TempOggPacket );
  if ( Ret >= 0 )
  {
```

This is the Vorbis header:

```
    memcpy( &VorbisStreamState, &OggStateTest,
      sizeof( OggStateTest ) );
    NumVorbisStreams = 1;
    continue;
  }
}
```

Since we need just the Theora stream information, ignore other codecs and drop the header:

```
        OGG_stream_clear( &OggStateTest );
    }
}
```

The preceding code basically just counts the number of streams, and we should be done by now. If there is still not enough streams, we will continue reading and checking secondary stream headers:

```
while((( NumTheoraStreams > 0 ) && ( NumTheoraStreams < 3 ))
  || (( NumVorbisStreams > 0 ) && ( NumVorbisStreams < 3 )))
{
  int Success = 0;
```

We will read all available packets and check if it is the beginning of a new Theora stream:

```
    while (( NumTheoraStreams > 0 ) &&
      ( NumTheoraStreams < 3 ) &&
      ( Success = OGG_stream_packetout( &TheoraStreamState,
        &TempOggPacket ) ) )
    {
      if ( Success < 0 ) return false;
      if ( !TH_decode_headerin( &TheoraInfo, &TheoraComment,
        &TheoraSetup, &TempOggPacket ) ) return false;
      ++NumTheoraStreams;
    }
```

In the same manner, we will look for the beginning of the next Vorbis stream:

```
    while ( NumVorbisStreams < 3 && ( Success =
      OGG_stream_packetout( &VorbisStreamState, &TempOggPacket ) ) )
    {
      if ( Success < 0 ) return false;
      if ( VORBIS_synthesis_headerin( &VorbisInfo,
        &VorbisComment, &TempOggPacket ) )
      return false;
      ++NumVorbisStreams;
    }
```

The final part of the long `while` (`!Done`) loop is to check for packets with the actual frame data, or to read more bytes from the stream if the next packet is not available:

```
if ( OGG_sync_pageout( &OggSyncState, &OggPage ) > 0 )
{
  if ( NumTheoraStreams > 0 )
  {
    OGG_stream_pagein( &TheoraStreamState, &OggPage );
  }
  if ( NumVorbisStreams > 0 )
  {
    OGG_stream_pagein( &VorbisStreamState, &OggPage );
  }
}
else
{
  char* Buffer = OGG_sync_buffer( &OggSyncState, 4096 );
  int BytesRead = (int)Stream_Read( Buffer, 4096 );
  OGG_sync_wrote( &OggSyncState, BytesRead );
  if ( BytesRead == 0 ) return false;
}
}
```

So far, we have found all stream headers and we are ready to initialize the Theora decoder. After doing so, we fetch the frame width and height:

```
TheoraDecoder = TH_decode_alloc( &TheoraInfo, TheoraSetup );
Width  = TheoraInfo.frame_width;
Height = TheoraInfo.frame_height;
return true;
}
```

In the end, we clear internal structures of the codec to avoid memory leaks:

```
void Theora_Cleanup()
{
  if ( TheoraDecoder )
  {
    TH_decode_free( TheoraDecoder );
    TH_setup_free( TheoraSetup );
    VORBIS_dsp_clear( &VorbisDSPState );
    VORBIS_block_clear( &VorbisBlock );
    OGG_stream_clear( &TheoraStreamState );
    TH_comment_clear( &TheoraComment );
```

```
        TH_info_clear( &TheoraInfo );
        OGG_stream_clear( &VorbisStreamState );
        VORBIS_comment_clear( &VorbisComment );
        VORBIS_info_clear( &VorbisInfo );
        OGG_sync_clear( &OggSyncState );
    }
}
```

This is it, we have read video parameters. We will return to audio and video decoding and playback in the following chapters, once we have some basic graphics and audio rendering capabilities in place.

The code more complex but similar to our example is heavily used inside the `LibTheoraPlayer` library source code available at `http://libtheoraplayer.cateia.com`.

In the example for this chapter, we will use capitalized function names to distinguish dynamic library usage and static linking. If you want to link the `ogg`, `vorbis`, and `theora` libraries statically, you may do so by renaming each `OGG` function prefix to `ogg`. That is it; just replace uppercase letters by lowercase ones.

For sample Theora video content, we will refer to the official website, `http://www.theora.org/content`, where you can download `.ogv` files.

OpenAL

OpenAL is a cross-platform audio API. It is designed for efficient rendering of multichannel three-dimensional positional audio and is widely used on desktop platforms in numerous gaming engines and applications. Many mobile platforms provide different audio APIs, for example, OpenSL ES is a strong player. However, when portability is at stake, we should select an API capable of running on all platforms required. OpenAL is implemented on Windows, Linux, OS X, Android, iOS, BlackBerry 10 and on many other platforms. On all these operating systems, except Windows and Android, OpenAL is a first person citizen with all libraries available in the system. On Windows, there is an implementation from Creative. On Android, we need to build the library ourself. We will use the Martins Mozeiko port `http://pielot.org/2010/12/14/openal-on-android/`. This library can be compiled for Android with minor tweaking of `Android.mk` and `Application.mk` files. Here is the `Android.mk` file:

```
TARGET_PLATFORM := android-19
LOCAL_PATH := $(call my-dir)
include $(CLEAR_VARS)
```

```
LOCAL_ARM_MODE := arm
LOCAL_MODULE := OpenAL
LOCAL_C_INCLUDES := $(LOCAL_PATH) $(LOCAL_PATH)/../include
$(LOCAL_PATH)/../OpenAL32/Include
LOCAL_SRC_FILES  := ../OpenAL32/alAuxEffectSlot.c \
                    ../OpenAL32/alBuffer.c \
                    ../OpenAL32/alDatabuffer.c \
                    ../OpenAL32/alEffect.c \
                    ../OpenAL32/alError.c \
                    ../OpenAL32/alExtension.c \
                    ../OpenAL32/alFilter.c \
                    ../OpenAL32/alListener.c \
                    ../OpenAL32/alSource.c \
                    ../OpenAL32/alState.c \
                    ../OpenAL32/alThunk.c \
                    ../Alc/ALc.c \
                    ../Alc/alcConfig.c \
                    ../Alc/alcEcho.c \
                    ../Alc/alcModulator.c \
                    ../Alc/alcReverb.c \
                    ../Alc/alcRing.c \
                    ../Alc/alcThread.c \
                    ../Alc/ALu.c \
                    ../Alc/android.c \
                    ../Alc/bs2b.c \
                    ../Alc/null.c
```

The -D definition are required for correct compilation:

```
GLOBAL_CFLAGS := -O3 -DAL_BUILD_LIBRARY -DAL_ALEXT_PROTOTYPES -
    DHAVE_ANDROID=1
```

And this if-block is a way to separate ARM and x86 compiler switches when you want to build an x86 version of the library for Android:

```
ifeq ($(TARGET_ARCH),x86)
  LOCAL_CFLAGS := $(GLOBAL_CFLAGS)
else
  LOCAL_CFLAGS := -mfpu=vfp -mfloat-abi=hard -mhard-float -fno-
    short-enums -D_NDK_MATH_NO_SOFTFP=1 $(GLOBAL_CFLAGS)
endif
include $(BUILD_STATIC_LIBRARY)
```

The `Application.mk` file is standard and looks as follows:

```
APP_OPTIM := release
APP_PLATFORM := android-19
APP_STL := gnustl_static
APP_CPPFLAGS += -frtti
APP_CPPFLAGS += -fexceptions
APP_CPPFLAGS += -DANDROID
APP_MODULES := OpenAL
APP_ABI := armeabi-v7a-hard x86
NDK_TOOLCHAIN_VERSION := clang
```

For your convenience, we have all the source code and configuration files in the `6_OpenAL` example. Furthermore, all the libraries we use in this book are precompiled for Android, you can find them in the `Libs.Android` folder within the book's source code bundle.

Linking the libraries to your application

There is one more thing left to discuss in this chapter before we proceed to further topics. Indeed, we learned how to build the libraries, but not how to link your Android application against them. For this purpose, we need to modify the `Android.mk` file of your application. Let's take a look at the `3_FreeImage_Example` sample and its `Application.mk`. It starts with the declaration of the prebuilt static library pointing to a binary file:

```
include $(CLEAR_VARS)
LOCAL_MODULE := libFreeImage
LOCAL_SRC_FILES
   :=../../../Libs.Android/libFreeImage.$(TARGET_ARCH_ABI).a
include $(PREBUILT_STATIC_LIBRARY)
```

Here, we use the `$(TARGET_ARCH_ABI)` variable in the path to transparently handle `armeabi-v7a-hard` and `x86` versions of the libraries. You can add yet more architectures with ease.

Once the library is declared, let's link the application against it. Take a look at the bottom of `Application.mk`:

```
LOCAL_STATIC_LIBRARIES += FreeImage
include $(BUILD_SHARED_LIBRARY)
```

The `LOCAL_STATIC_LIBRARIES` variable contains all the required libraries. The prefix `lib` can be omitted for your convenience.

Summary

In this chapter, we learned how to deal with precompiled static libraries on Android, same also applies to OS X and Linux, and how to do dynamic linking on Windows without breaking multiplatform capabilities of our code. We learned how to build `libcurl` and `OpenSSL`, so you can access SSL connections from your C++ code. A couple of examples for FreeImage and FreeType shows how to load and save images with raster fonts. The example with libtheora was quite comprehensive; however, the result was modest, we just read meta-information from a video file. OpenAL will be used as a backbone of our audio subsystem.

3
Networking

In this chapter, we will learn how to deal with network-related functionality from the native C/C++ code. Networking tasks are asynchronous by nature and unpredictable in terms of timing. Even when the underlying connection is established using the TCP protocol, there is no guarantee on the delivery time, and nothing prevents the applications from freezing while waiting for the data. In the Android SDK, this is hidden from a developer by a myriad of classes and facilities. In Android NDK, *au contraire*, one has the responsibility to overcome these difficulties without assistance from any platform-specific helpers. To develop responsive and safe applications, a number of problems must be solved: we need to be in full control of the download process, we have to limit the downloaded data size, and gracefully handle the errors that occur. Without delving into the details of the HTTP and SSL protocols implementation, we will use the libcurl and OpenSSL libraries, and concentrate on higher-level tasks related to application development. However, we will take a closer look at implementing basic asynchronous mechanisms in a portable way. The first few examples of this chapter are desktop-only and their purpose is to show how a multiplatform synchronization primitives can be implemented. However, at the end of this chapter, we will see how to put all these pieces together into a mobile application.

Intrusive smart pointers

Tracking all native memory allocations in a multithreaded environment is notoriously difficult process, especially when it comes to passing objects ownership between different threads. In C++, memory management can be automated using smart pointers. The standard `std::shared_ptr` class is a good place to start with. However, we want to focus on more interesting and lightweight techniques. We will not use the Boost library either, since we really want to stay lean when it comes to compile times.

The recent versions of the Android NDK support the C++ 11 Standard Library in full. If you feel more comfortable with std::shared_ptr or intrusive pointers from Boost, feel free to use smart pointers from those libraries.

In an intrusive smart pointer, as the name suggests, a reference counter is embedded into the object. The simplest way to do this is by inheriting from the following base class:

```
class iIntrusiveCounter
{
private:
  std::atomic<long> m_RefCounter;
public:
  iIntrusiveCounter( ) : m_RefCounter( 0 ) {}
  virtual ~iIntrusiveCounter( ) {}
  long GetReferenceCounter( ) const volatile
  { return m_RefCounter; }
```

It uses a standard atomic variable to hold the value of the counter. Before C++ 11 Standard Library was widely adopted, a portable implementation of an atomic counter required platform-specific atomic operations to be used, either POSIX or Windows. Nowadays, it is possible to write clean code for all platforms using C++ 11; for Android, Windows, Linux, OS X, iOS, and even for BlackBerry 10, if you wish to do so. Here is how we can increment the counter:

```
void IncRefCount( )
{
  m_RefCounter.fetch_add( 1, std::memory_order_relaxed );
}
```

It is absolutely possible to use the ++ operator instead of fetch_add(). However, incrementing an atomic integer variable this way is required by the compiler to be sequentially consistent and may cause redundant memory barriers to be inserted in the generated assembly. We do not use the incremented value for any decision making, hence the memory barriers here are unnecessary and memory ordering can be relaxed, only atomicity of the variable is required. This is what fetch_add() does with the std::memory_order_relaxed flag, leading to a faster code on some non-x86 platforms. Decrement is trickier. Indeed, we have to decide when to remove the object, and do so only when the reference counter is decremented to zero.

Here is the code to do it right:

```
void DecRefCount()
{
   if ( m_RefCounter.fetch_sub( 1, std::memory_order_release ) == 1 )
   {
```

The `std::memory_order_release` flag means an operation on a memory location requires all previous memory writes to become visible to all threads that do an acquire operation on the same location. After entering the `if` block, we will do the acquire operation by inserting an appropriate memory barrier:

```
std::atomic_thread_fence( std::memory_order_acquire );
```

Only after this point, we can now allow the object to commit suicide:

```
delete this;
      }
   }
};
```

The `delete this` idiom is explained at https://isocpp.org/wiki/faq/freestore-mgmt#delete-this.

 The `iIntrusiveCounter` class is the core of our reference-counting mechanism. The code may look very simple; however, the logic behind this implementation is much more complicated than it seems. Refer to the *C++ and Beyond 2012: Herb Sutter - atomic<> Weapons, 1 of 2* talk by Herb Sutter for all the elaborate details:

http://channel9.msdn.com/Shows/Going+Deep/Cpp-and-Beyond-2012-Herb-Sutter-atomic-Weapons-1-of-2

http://channel9.msdn.com/Shows/Going+Deep/Cpp-and-Beyond-2012-Herb-Sutter-atomic-Weapons-2-of-2

Now, we can implement a lightweight RAII generic smart pointer class, which utilizes our freshly written counter base class:

```
template <class T> class clPtr
{
public:
   /// default constructor
   clPtr(): FObject( 0 ) {}
```

```
/// copy constructor
clPtr( const clPtr& Ptr ): FObject( Ptr.FObject )
{
  LPtr::IncRef( FObject );
}
```

Here, a copy constructor does not call the `FObject->IncRefCount()` method directly. Instead, it invokes a helper function `LPtr::IncRef()`, which accepts `void*` and passes the object as a parameter to that function. This is done to allow usage of our intrusive smart pointer with classes that were declared but not yet defined:

```
/// move constructor
clPtr( clPtr&& Ptr :): FObject( Ptr.FObject )
{
  Ptr.FObject = nullptr;
}
template <typename U> clPtr( const clPtr<U>& Ptr )): FObject(
  Ptr.GetInternalPtr() )
{
  LPtr::IncRef( FObject );
}
```

The implicit constructor from `T*` is useful:

```
clPtr( T* const Object ): FObject( Object )
{
  LPtr::IncRef( FObject );
}
```

Similar to the constructor, the destructor uses a helper function to decrement the reference counter:

```
~clPtr()
{
  LPtr::DecRef( FObject );
}
```

A couple of named helper functions can be of use to check the state of the smart pointer:

```
/// check consistency
inline bool IsValid() const
{
  return FObject != nullptr;
```

```
    }
    inline bool IsNull() const
    {
      return FObject == nullptr;
    }
```

Assignment is quite slow compared to other methods:

```
    /// assignment of clPtr
    clPtr& operator = ( const clPtr& Ptr )
    {
      T* Temp = FObject;
      FObject = Ptr.FObject;
      LPtr::IncRef( Ptr.FObject );
      LPtr::DecRef( Temp );
      return *this;
    }
```

But not the move assignment operator:

```
    clPtr& operator = ( clPtr&& Ptr )
    {
      FObject = Ptr.FObject;
      Ptr.FObject = nullptr;
      return *this;
    }
```

The -> operator is essential for every smart pointer class:

```
    inline T* operator -> () const
    {
      return FObject;
    }
```

And here is a bit tricky one: an automatic type conversion operator to an instance of the private class clProtector:

```
    inline operator clProtector* () const
    {
      if ( !FObject ) return nullptr;
      static clProtector Protector;
      return &Protector;
    }
```

This type conversion is used to allow safe null-pointer checks as in `if ( clPtr )` to be possible. It is safe because you cannot do anything with the resulting pointer. The inner private class `clProtector` does not implement the `delete()` operator, hence, using it will produce a compilation error:

```
private:
  class clProtector
  {
private:
    void operator delete( void* ) = delete;
  };
```

> The source code bundle of this book does not use the = delete C++ 11 notation for the deleted function, but just leaves it unimplemented. This is done for the sake of compatibility with older compilers. If you target the most recent versions of GCC/Clang and Visual Studio, going with = delete will be just fine.

Let's return to our `clPtr` class. Unfortunately, the standard `dynamic_cast<>` operator cannot be used in the original way, so we need to make a substitution:

```
public:
  /// cast
  template <typename U> inline clPtr<U> DynamicCast() const
  {
    return clPtr<U>( dynamic_cast<U*>( FObject ) );
  }
```

This is the only thing where our smart pointer will be different in syntax from raw pointers. Also, we need a set of comparison operators to make our class more useful within different containers:

```
template <typename U> inline
  bool operator == ( const clPtr<U>&Ptr1 ) const
{
  return FObject == Ptr1.GetInternalPtr();
}
template <typename U> inline
  bool operator == ( const U* Ptr1 )const
{
  return FObject == Ptr1;
}
```

```
template <typename U> inline
  bool operator != ( const clPtr<U>&Ptr1 ) const
{
  return FObject != Ptr1.GetInternalPtr();
}
```

Here is a function to abridge the smart pointer with APIs accepting raw pointers. The conversion to the underling `T*` type should be explicit:

```
inline T* GetInternalPtr() const
{
  return FObject;
}
```

Some helper functions may be useful when dealing with low-level pointer fuss. Drop the object, do not deallocate it:

```
inline void Drop()
{
  FObject = nullptr;
}
```

Clear the object, decrement the reference counter, similar to assigning `nullptr` to it:

```
inline void Clear()
{
  *this = clPtr<T>();
}
```

The last but not least, the pointer itself:

```
private:
  T* FObject;
};
```

Henceforth, our portable intrusive smart pointer is self-contained and can be used in real applications. There is one more thing left to be done, a kind of syntactic sugar. It is typical of C++ 11 to use the `auto` keyword, so one can write the type name in the expression only once. However, the following instantiation will not work because the deduced type of `p` will be `clSomeObject*` when we wanted it to be `clPtr< clSomeObject>`:

```
auto p = new clSomeObject( a, b, c );
```

With standard shared pointers, this is solved using a `std::make_shared()` template helper function, which returns the proper type (and does some useful optimization of the counter storage behind the scenes):

```
auto p = std::make_shared<clSomeObject>( a, b, c );
```

Here, the deduced type of p is `std::shared_ptr<clSomeObject>`, which ultimately matches our expectations. We can create a similar helper using the perfect forwarding mechanism provided by C++ 11 and the `std::forward()` function:

```
template< class T, class... Args > clPtr<T> make_intrusive(
  Args&&... args )
{
  return clPtr<T>( new T( std::forward<Args>( args )... ) );
}
```

The usage is C++11-stylish and natural:

```
auto p = make_intrusive<clSomeObject>( a, b, c );
```

The complete source code for the smart pointer can be found in the `1_IntrusivePtr` example. Now, we can proceed further and use this class as the cornerstone of our multithreaded memory management.

Portable multithreading primitives

The long-awaited `std::thread` from the C++11 standard is not (yet) available in MinGW toolchain at the time of writing, and it does not possess capabilities necessary to adjust thread priorities, which is important for networking. So, we implement a simple class `iThread` with the virtual method `Run()` to allow portable multithreading in our code:

```
class iThread
{
```

An internal `LPriority` enumeration defines thread priority classes:

```
public:
  enum LPriority
  {
    Priority_Idle         = 0,
    Priority_Lowest       = 1,
    Priority_Low          = 2,
    Priority_Normal       = 3,
```

```
    Priority_High        = 4,
    Priority_Highest     = 5,
    Priority_TimeCritical = 6
};
```

The code for constructor and destructor is simple:

```
iThread(): FThreadHandle( 0 ), FPendingExit( false )
{}
virtual ~iThread()
{}
```

The `Start()` method creates an OS-specific thread handle and starts execution. In all of the samples for this book, we do not need to postpone thread execution; we just call `_beginthreadex()` and `pthread_create()` system routines with default parameters. The `EntryPoint()` method is defined later:

```
void Start()
{
  void* ThreadParam = reinterpret_cast<void*>( this );
  #ifdef _WIN32
    unsigned int ThreadID = 0;
    FThreadHandle = ( uintptr_t )_beginthreadex(
      nullptr, 0, &EntryPoint, ThreadParam, 0, &ThreadID );
  #else
    pthread_create( &FThreadHandle, nullptr,
      EntryPoint, ThreadParam );
    pthread_detach( FThreadHandle );
  #endif
}
```

The system-dependent thread handle and the Boolean atomic variable, which indicates whether this thread should stop its execution, are declared in the private section of the class:

```
private:
  thread_handle_t FThreadHandle;
  std::atomic<bool> FpendingExit;
```

The native threading API supports only C functions, therefore we have to declare a static wrapper method `EntryPoint()`, which converts its `void*` parameter to `iThread` and calls the `Run()` method of the class. The calling convention and result type for the thread function differs on POSIX and Windows:

```
#ifdef _WIN32
  #define THREAD_CALL unsigned int __stdcall
#else
  #define THREAD_CALL void*
#endif
  static THREAD_CALL EntryPoint( void* Ptr );
```

The protected section defines the `Run()` and `NotifyExit()` virtual methods, which are overridden in subclasses. The `GetHandle()` method allows subclasses to access the platform-specific thread handle:

```
protected:
  virtual void Run() = 0;
  virtual void NotifyExit() {};
  thread_handle_t GetHandle() { return FThreadHandle; }
```

To stop the thread, we will raise the `FPendingExit` flag and call the `NotifyExit()` method to inform the thread owner. An optional `Wait` parameter forces the method to wait for actual termination of the thread:

```
void Exit( bool Wait )
{
  FPendingExit = true;
  NotifyExit();
  if ( !Wait ) { return; }
```

We must ensure `Exit()` is not called from the `Run()` method of the same thread to avoid deadlocks, so we will call `GetCurrentThread()` and compare the result with our own handle:

```
if ( GetCurrentThread() != FThreadHandle )
{
```

For Windows, we will mimic the `join` operation by calling `WaitForSingleObject()` and then terminating the thread via `CloseHandle()`:

```
#ifdef _WIN32
  WaitForSingleObject(( HANDLE )FThreadHandle, INFINITE );
  CloseHandle( ( HANDLE )FThreadHandle );
```

```
    #else
      pthread_join( FThreadHandle, nullptr );
    #endif
  }
}
```

The `GetCurrentThread()` method on Android is implemented slightly different from a typical POSIX version. Hence, this method contains a triple `#ifdef` clause:

```
native_thread_handle_t iThread::GetCurrentThread()
{
  #if defined( _WIN32)
    return GetCurrentThreadId();
  #elif defined( ANDROID )
    return gettid();
  #else
    return pthread_self();
  #endif
}
```

The `EntryPoint()` method is the glue that ties together our object-oriented `iThread` wrapper class and the platform-specific C-style threading API:

```
THREAD_CALL iThread::EntryPoint( void* Ptr )
{
  iThread* Thread = reinterpret_cast<iThread*>( Ptr );
  if ( Thread )
  {
    Thread->Run();
  }
  #ifdef _WIN32
    _endthreadex( 0 );
    return 0;
  #else
    pthread_exit( 0 );
  return nullptr;
  #endif
}
```

The finishing touch is the `SetPriority()` method that is used to control the amount of CPU time given for the thread. For Windows, the main part of the method is the conversion of our `LPriority` enumeration to the numerical value defined in the `windows.h` header file:

```
void iThread::SetPriority( LPriority Priority )
{
  #ifdef _WIN32
    int P = THREAD_PRIORITY_IDLE;
    switch(Priority)
    {
      case Priority_Lowest:
        P = THREAD_PRIORITY_LOWEST; break;
      case Priority_Low:
        P = THREAD_PRIORITY_BELOW_NORMAL; break;
      case Priority_Normal:
        P = THREAD_PRIORITY_NORMAL; break;
      case Priority_High:
        P = THREAD_PRIORITY_ABOVE_NORMAL; break;
      case Priority_Highest:
        P = THREAD_PRIORITY_HIGHEST; break;
      case Priority_TimeCritical:
        P = THREAD_PRIORITY_TIME_CRITICAL; break;
    }
    SetThreadPriority( ( HANDLE )FThreadHandle, P );
  #else
```

For POSIX, we rescale our priority value to the integer number between minimum and maximum priorities available in the OS:

```
    int SchedPolicy = SCHED_OTHER;
    int MaxP = sched_get_priority_max( SchedPolicy );
    int MinP = sched_get_priority_min( SchedPolicy );
    sched_param SchedParam;
    SchedParam.sched_priority = MinP + (MaxP - MinP) /
      (Priority_TimeCritical - Priority + 1);
    pthread_setschedparam( FThreadHandle, SchedPolicy, &SchedParam
  );
  #endif
}
```

Now, we can use the `iThread` class to construct more useful higher-level threading primitives. For cross-platform lightweight implementation of the `std::mutex`-like object, we will use the TinyThread library by Marcus Geelnard, which is available for download at `http://tinythreadpp.bitsnbites.eu`. However, feel free to use the standard mutex if you do not have to be compatible with legacy compilers.

Let's proceed with the task queues.

Task queues

To process a logical piece of work, we will declare the `iTask` class with the `Run()` method, which can perform a time-consuming operation. The declaration of the class is somewhat visually similar to `iThread`. However, its instances implement some reasonably short operation and may be executed in different threads:

```
class iTask: public iIntrusiveCounter
{
public:
  iTask()
  : FIsPendingExit( false )
  , FTaskID( 0 )
  , FPriority( 0 )
  {};
```

The pure virtual method `Run()` should be overridden in subclasses to do the actual work:

```
virtual void Run() = 0;
```

The following methods optionally cancel the task and are similar to the ones in the `iThread` class. Their purpose is to signal the hosting thread that this task should be cancelled:

```
virtual void Exit()
{
  FIsPendingExit = true;
}
virtual bool IsPendingExit() const volatile
{
  return FIsPendingExit;
}
```

The `GetTaskID()` and `SetTaskID()` methods access the internal unique identifier of the task, which is used to cancel execution:

```
virtual void SetTaskID( size_t ID )
{ FTaskID = ID; };
virtual size_t GetTaskID() const
{ return FTaskID; };
```

The `GetPriority()` and `SetPriority()` methods are used by the task scheduler to determine the order in which the tasks are executed:

```
virtual void SetPriority( int P )
{
  FPriority = P;
};
virtual int GetPriority() const
{
  return FPriority;
};
```

The private part of the class contains an atomic exit flag, the task ID value, and task priority:

```
private:
  std::atomic<bool> FIsPendingExit;
  size_t FTaskID;
  int FPriority;
};
```

The management of the tasks is done by the `clWorkerThread` class. Basically, it is a collection of `iTask` instances, which is fed using the `AddTask()` method. The private part of the class contains `std::list` of `iTask`s and a few synchronization primitives:

```
class clWorkerThread: public iThread
{
private:
  std::list< clPtr<iTask> >   FPendingTasks;
  clPtr<iTask>                FCurrentTask;
  mutable tthread::mutex      FTasksMutex;
  tthread::condition_variable FCondition;
```

The `FCurrentTask` field is used internally to keep track of the task, which is in progress. The `FTasksMutex` field is a mutex to ensure thread-safe access to `FPendingTasks`. The `FCondition` conditional variable is used to signal the availability of tasks in the list.

The `AddTask()` method inserts a new task to the list and notifies the `Run` method about the task availability:

```
virtual void   AddTask( const clPtr<iTask>& Task )
{
  tthread::lock_guard<tthread::mutex> Lock( FTasksMutex );
  FPendingTasks.push_back( Task );
  FCondition.notify_all();
}
```

To check whether there are unfinished tasks, we will define the `GetQueueSize()` method. The method uses the `std::list.size()` and increments the returned value if there is an active task currently running:

```
virtual size_t GetQueueSize() const
{
  tthread::lock_guard<tthread::mutex> Lock( FTasksMutex );
  return FPendingTasks.size() + ( FCurrentTask ? 1 : 0 );
}
```

There is the `CancelTask()` method to cancel a single task and the `CancelAll()` method to cancel all of the tasks at once:

```
virtual bool   CancelTask( size_t ID )
{
  if ( !ID ) { return false; }
  tthread::lock_guard<tthread::mutex> Lock( FTasksMutex );
```

First, we check whether a task is running and its ID matched with the one we want to cancel:

```
if ( FCurrentTask && FCurrentTask->GetTaskID() == ID )
  FCurrentTask->Exit();
```

Then, we will iterate the list of tasks and request exit for the one with the given ID, removing them from the list of pending tasks. This can be done using a simple lambda:

```
FPendingTasks.remove_if(
  [ID]( const clPtr<iTask> T )
  {
    if ( T->GetTaskID() == ID )
    {
      T->Exit();
      return true;
    }
    return false;
  }
);
```

Finally, we notify everyone about the list change:

```
FCondition.notify_all();
return true;
}
```

The `CancelAll()` method is much simpler. The task list is iterated and every item is requested to terminate; after this, the container is cleared and a notification is sent:

```
virtual void CancelAll()
{
  tthread::lock_guard<tthread::mutex> Lock( FTasksMutex );
  if ( FcurrentTask )
  {
    FcurrentTask->Exit();
  }
  for ( auto& Task: FpendingTasks )
  {
    Task->Exit();
  }
  FpendingTasks.clear();
  Fcondition.notify_all();
}
```

The main work is done in the `Run()` method, which waits for a next task to arrive and executes it:

```
virtual void Run()
{
```

The outer loop checks whether we need to stop this worker thread using the `iThread::IsPendingExit()` routine:

```
while ( !IsPendingExit() )
{
```

The `ExtractTask()` method extracts the next task from the list. It waits on the conditional variable until the task is actually available:

```
FCurrentTask = ExtractTask();
```

If the task is valid and its cancellation is not requested, we can start the task:

```
if ( FCurrentTask &&
  !FCurrentTask->IsPendingExit())
FCurrentTask->Run();
```

After the task has completed its work, we will clear the state to ensure the correct `GetQueueSize()` operation:

```
FCurrentTask = nullptr;
  }
}
```

The `ExtractTask()` method implements a thread-safe linear search in the `FPendingTasks` list to select the `iTask` instance with the highest priority:

```
clPtr<iTask> ExtractTask()
{
  tthread::lock_guard<tthread::mutex> Lock( FTasksMutex );
```

To avoid doing spinlock and burning CPU cycles, the conditional variable is checked:

```
while ( FPendingTasks.empty() && !IsPendingExit() )
  FCondition.wait( FTasksMutex );
```

If the list is empty, the empty smartpointer is returned:

```
if ( FPendingTasks.empty() )
  return clPtr<iTask>();
```

The `Best` variable stores the selected task to be executed:

```
auto Best = FPendingTasks.begin();
```

Iterating over the `FPendingTask` list and comparing the priority value to the one in the `Best` variable, we will select the task:

```
for ( auto& Task : FPendingTasks )
{
  if ( Task->GetPriority() >
     ( *Best )->GetPriority() ) *Best = Task;
}
```

Finally, we will erase the selected task from the container and return the result. A temporary variable is needed to ensure our smartpointer does not decrement the reference count to zero:

```
clPtr<iTask> Result = *Best;
FPendingTasks.erase( Best );
Return Result;
}
```

Now, we have the class to handle asynchronous tasks. There is one more crucial thing to be done before we can proceed with the actual asynchronous networking—asynchronous callbacks.

Message pumps and asynchronous callbacks

In the previous section, we defined `clWorkerThread` and `iTask` classes that allow us to execute lengthy operations outside of the UI thread in C++ code. The final thing we need to organize a responsive interface is the ability to pass around events across different threads. To do this, we need a callable interface, which can encapsulate the parameters passed to a method, and a thread-safe mechanism to pass such capsules across the threads.

A nice candidate for such a capsule is `std::packaged_task`, which is not supported in the most recent MinGW toolchain. Therefore, we will define our own lightweight reference-counted abstract class `iAsyncCapsule`, which implements a single method, `Invoke()`:

```
class iAsyncCapsule: public iIntrusiveCounter
{
public:
  virtual void Invoke() = 0;
};
```

We call a prioritized collection of `iAsyncCapsule` instances wrapped in `clPtr` an *asynchronous queue*. The `clAsyncQueue` class implements the `DemultiplexEvents()` method, which will be called in the thread that does the processing of incoming events.

 This is called the reactor pattern. It's documentation can be found at http://en.wikipedia.org/wiki/Reactor_pattern.

The demultiplexing consists of invoking all the accumulated `iAsyncCapsule`s that are added from the other threads by the `EnqueueCapsule()` method. Both methods should be and are thread-safe. However, `DemultiplexEvents()` is not reentrant in the sense that no two threads should invoke `DemultiplexEvents()` on the same object. This limitation is a part of performance optimization, which we will see further. We use two containers of `iAsyncCapsule`s and switch between them at each call of `DemultiplexEvents()`. This allows faster `EnqueueCapsule()` execution because we don't have to copy the contents of the queue to ensure thread-safety. Otherwise, a copy would be necessary because we should not call `Invoke()` while the mutex is locked.

The private section of the class contains the index of the current queue in use `FCurrentQueue`, two containers of `iAsyncCapsules`, a pointer to the current queue, and a mutex to prevent simultaneous access to the `FAsyncQueues` array:

```
class clAsyncQueue
{
private:
  using CallQueue = std::vector< clPtr<iAsyncCapsule> >;
  size_t FCurrentQueue;
  std::array<CallQueue, 2> FAsyncQueues;
  /// switched for shared non-locked access
  CallQueue* FAsyncQueue;
  tthread::mutex FDemultiplexerMutex;
```

Constructor initializes the current queue pointer and the index:

```
public:
  clAsyncQueue()
  : FDemultiplexerMutex()
  , FCurrentQueue( 0 )
  , FAsyncQueues()
  , FAsyncQueue( &FAsyncQueues[0] )
  {}
```

The `EnqueueCapsule()` method is similar to `WorkerThread::AddTask()`. First, we create a scoped `lock_guard` object, and then call `push_back()` to enqueue the `iAsyncCapsule` object:

```
virtual void EnqueueCapsule(
  const clPtr<iAsyncCapsule>& Capsule )
{
  tthread::lock_guard<tthread::mutex>
    Lock( FDemultiplexerMutex );
  FAsyncQueue->push_back( Capsule );
}
```

The `DemultiplexEvents()` method saves the reference to the current queue:

```
virtual void DemultiplexEvents()
{
```

`DemultiplexEvents()` is designed to run only on one thread. No locking is needed at this point:

```
CallQueue& LocalQueue = FAsyncQueues[ FCurrentQueue ];
```

Then, the current queue pointer is swapped. This is an atomic operation, so we use lock the mutex to prevent access to the FAsyncQueue pointer and the index:

```
    {
      tthread::lock_guard<tthread::mutex>
        Lock( FDemultiplexerMutex );
      FCurrentQueue = ( FCurrentQueue + 1 ) % 2;
      FAsyncQueue = &FAsyncQueues[ FCurrentQueue ];
    }
```

Finally, each and every iAsyncCapsule in the current queue is invoked and the LocalQueue is cleared:

```
      for ( auto& i: LocalQueue ) i->Invoke();
      LocalQueue.clear();
    }
  };
```

The typical usage scenario is posting callbacks from one thread to another. A small sample considered here uses the clResponseThread class with an endless loop as the main thread:

```
    class clResponseThread: public iThread, public clAsyncQueue
    {
    public:
      virtual void Run()
      {
        for (;;) DemultiplexEvents();
      }
    };
```

The sample clRequestThread class is producing events twice per second:

```
    class clRequestThread: public iThread
    {
    public:
      explicit clRequestThread( clAsyncQueue* Target )
      : FTarget(Target)
      {}
      virtual void Run()
      {
        int id = 0;
        for (;;)
        {
          FTarget->EnqueueCapsule( make_intrusive<clTestCall>( id++ ) );
```

```
      OS_Sleep( 500 );
    }
  }
private:
  clAsyncQueue* FTarget;
};
```

The test call just prints out a message with the `clTestCall` ID:

```
class clTestCall: public iAsyncCapsule
{
private:
  int id;
public:
  explicit clTestCall( int i ): id(i) {}
  virtual void Invoke()
  {
    std::cout "Test " << id << std::endl;
  }
};
```

In the `main()` function, we create both threads and start an infinite loop:

```
clResponseThread Responder;
clRequestThread Requester( &Responder );
Responder.Start();
Requester.Start();
for (;;) {}
```

In the following section, we will use a similar approach to inform the main thread of the download result. The `clResponseThread` class becomes the UI thread and `clRequestThread` is a `WorkerThread` method where each of executed download tasks fires an event once the download is finished.

Asynchronous networking with libcurl

The simplistic usage of libcurl was shown in *Chapter 2, Native Libraries*. Now, we extend the code using the previously mentioned multithreading primitives to allow asynchronous downloads.

The `clDownloadTask` class introduced here keeps track of the download process and invokes a callback when the process completes:

```
class clDownloadTask: public iTask
{
public:
```

The constructor accepts URL of the resource to download, a unique task identifier, a callback, and a pointer to the instance of the downer:

```
clDownloadTask( const std::string& URL,
  size_t TaskID,
  const clPtr<clDownloadCompleteCallback>& CB,
  clDownloader* Downloader );
```

We will focus on the `Run()`, `Progress()`, and `InvokeCallback()` methods, as they form the main logic of this class:

```
virtual void Run() override;
private:
  void Progress( double TotalToDownload,
    double NowDownloaded,
    double TotalToUpload,
    double NowUploaded );
  void InvokeCallback();
};
```

The `Run()` method runs on the download thread; it initializes and performs the actual downloading of the resource using libcurl:

```
void clDownloadTask::Run()
{
```

This hard reference is required to prevent external destruction of the task if it is cancelled:

```
clPtr<clDownloadTask> Guard( this );
CURL* Curl = curl_easy_init_P();
```

The initialization of libcurl goes here. All possible parameters can be found in the official documentation at http://curl.haxx.se/libcurl/c/curl_easy_setopt. html:

```
curl_easy_setopt_P( Curl, CURLOPT_URL, FURL.c_str() );
curl_easy_setopt_P( Curl, CURLOPT_FOLLOWLOCATION, 1 );
curl_easy_setopt_P( Curl, CURLOPT_NOPROGRESS, false );
curl_easy_setopt_P( Curl, CURLOPT_FAILONERROR, true );
curl_easy_setopt_P( Curl, CURLOPT_MAXCONNECTS, 10 );
curl_easy_setopt_P( Curl, CURLOPT_MAXFILESIZE, DownloadSizeLimit );
curl_easy_setopt_P( Curl, CURLOPT_WRITEFUNCTION,
  &MemoryCallback );
```

```
curl_easy_setopt_P( Curl, CURLOPT_WRITEDATA, this );
curl_easy_setopt_P( Curl, CURLOPT_PROGRESSFUNCTION,
  &ProgressCallback );
curl_easy_setopt_P( Curl, CURLOPT_PROGRESSDATA, this );
```

The following line sets the number of seconds to wait while trying to connect. Use the value of zero to wait indefinitely:

```
curl_easy_setopt_P( Curl, CURLOPT_CONNECTTIMEOUT, 30 );
```

Here we set the maximal number of seconds to allow libcurl functions to execute:

```
curl_easy_setopt_P( Curl, CURLOPT_TIMEOUT, 600 );
```

Disable verification of certificates by OpenSSL, this will enable access to sites with self-signed certificates. However, you may want to remove this mode in the production code to reduce the possibility of man-in-the-middle attacks:

```
curl_easy_setopt_P( Curl, CURLOPT_SSL_VERIFYPEER, 0 );
curl_easy_setopt_P( Curl, CURLOPT_SSL_VERIFYHOST, 0 );
curl_easy_setopt_P( Curl, CURLOPT_HTTPGET, 1 );
```

When negotiating an SSL connection, the server sends a certificate indicating its identity. Curl verifies whether the certificate is authentic—that is, that you can trust that the server is who the certificate says it is. This trust is based on a chain of digital signatures, rooted in certification authority (CA) certificates you supply.

You can find the documentation at the following URLs:

http://curl.haxx.se/libcurl/c/CURLOPT_SSL_VERIFYPEER.html

http://curl.haxx.se/libcurl/c/CURLOPT_SSL_VERIFYHOST.html

Perform the actual download:

```
FCurlCode = curl_easy_perform_P( Curl );
curl_easy_getinfo_P( Curl, CURLINFO_RESPONSE_CODE, &FRespCode );
curl_easy_cleanup_P( Curl );
```

Let the downloader deal with the results of this task. We will follow this code shortly:

```
        if ( FDownloader ) { FDownloader->CompleteTask( this ); }
    }
```

The private `InvokeCallback()` member function is accessible from the friend class `clDownloader`:

```
    void clDownloadTask::InvokeCallback()
    {
        tthread::lock_guard<tthread::mutex> Lock( FExitingMutex );
```

Per se, this is just a `FCallback->Invoke()` invocation augmented with two runtime checks. The first one checks whether the task is not cancelled:

```
    if ( !IsPendingExit() )
    {
      if ( FCurlCode != 0 )
      {
        FResult = nullptr;
      }
```

The second one checks the availability of the callback and prepared all the parameters:

```
    if ( FCallback )
    {
      FCallback->FTaskID = GetTaskID();
      FCallback->FResult = FResult;
      FCallback->FTask = clPtr<clDownloadTask>( this );
      FCallback->FCurlCode = FCurlCode;
      FCallback->Invoke();
      FCallback = nullptr;
    }
  }
}
```

It is important to notice the callback invocation is done while the mutex is locked. Doing so is essential to ensure the correct cancellation behavior. However, `InvokeCallback()` is not called directly from `clDownloadTask`. Instead, there is an indirection via the `FDownloader->CompleteTask( this )` called from the `Run()` method. Let's see what is inside of it and look at the guts of the `clDownloader` class:

```
class clDownloader: public iIntrusiveCounter
{
public:
   explicit clDownloader( const clPtr<clAsyncQueue>& Queue );
   virtual ~clDownloader();
```

This method is the most important part of our public downloading API:

```
   virtual clPtr<clDownloadTask> DownloadURL(
     const std::string& URL, size_t TaskID,
     const clPtr<clDownloadCompleteCallback>& CB );
   virtual bool CancelLoad( size_t TaskID );
   virtual void CancelAll();
   virtual size_t GetNumDownloads() const;
```

And this one handles the indirection:

```
private:
   void CompleteTask( clPtr<clDownloadTask> Task );
   friend class clDownloadTask;
```

This is the thread where `clDownloadTask`s are run:

```
   clPtr<clWorkerThread> FDownloadThread;
```

An external event queue is initialized using the constructor parameter:

```
   clPtr<clAsyncQueue> FEventQueue;
};
```

Though, the `DownloadURL()` method is crucial, its implementation is surprisingly simple:

```
clPtr<clDownloadTask> DownloadURL( const std::string& URL,
   size_t TaskID,
   const clPtr<clDownloadCompleteCallback>& CB )
{
   if ( !TaskID || !CB ) { return clPtr<clDownloadTask>(); }
```

```
    auto Task = make_intrusive<clDownloadTask>(
        URL, TaskID, CB, this );
    FDownloadThread->AddTask( Task );
    return Task;
}
```

Indeed, all the hard work is done inside the aforementioned method
`clDownloadTask::Run()`. Here, we just enqueued a newly constructed task into
the worker thread. The most interesting thing happens inside `CompleteTask()`:

```
void clDownloader::CompleteTask( clPtr<clDownloadTask> Task )
{
    if ( !Task->IsPendingExit() )
    {
        if ( FEventQueue )
        {
```

Here, a callback wrapper is inserted into the event queue:

```
            FEventQueue->EnqueueCapsule(
                make_intrusive<clCallbackWrapper>(Task) );
        }
    }
}
```

The helper class calls the very `FTask->InvokeCallback()` method. Remember,
the method is invoked on the right thread, it was dispatched by the event queue:

```
class clCallbackWrapper: public iAsyncCapsule
{
public:
    explicit clCallbackWrapper(
        const clPtr<clDownloadTask> T ):FTask(T) {}
    virtual void Invoke() override
    {
        FTask->InvokeCallback();
    }
private:
    clPtr<clDownloadTask> FTask;
};
```

The usage sample can be found in the 3_Downloader folder of the source code
bundle. It is as simple as this snippet:

```
int main()
{
    Curl_Load();
```

This queue will handle download results:

```
auto Events = make_intrusive<clAsyncQueue>();
auto Downloader = make_intrusive<clDownloader>( Events );
clPtr<clDownloadTask> Task = Downloader->DownloadURL(
    http://downloads.sourceforge.net/freeimage/FreeImage3160.zip,
    1, make_intrusive<clTestCallback>() );
while ( !g_ShouldExit ) { Events->DemultiplexEvents(); }
return 0;
}
```

The `clTestCallback` class prints downloading progress and saves results to a file, `.zip` in our example.

> We use the `LUrlParser` library to extract the file name from a given URL, `https://github.com/corporateshark/LUrlParser`.

The sample code can be compiled using MinGW by typing `make all`. The same code can be run on Android without changes using the compiled Curl library from *Chapter 2, Native Libraries*. We suggest that you experiment with this code on Android and play with some downloads directly from the C++ code.

Android licensing in native applications

The major part of this chapter has been dedicated to low-level networking capabilities in C++, which are crucial to writing multiplatform code. However, it would not be fair to omit some Android-specific things from this chapter. Let's go with licensing mechanism and learn how to move it into the C++ code. For this one, we will need to interact with Java code heavily, since all the licensing facilities are Java-only.

> Here, we assume that you are already familiar with how to do the license checking in Java. The official Google documentation can be found here:
>
> `http://developer.android.com/google/play/licensing/setting-up.html`
>
> `http://developer.android.com/google/play/licensing/adding-licensing.html`

The source code of this sample is located in the 4_Licensing folder. First, let's define the basic constants, the values should match those from the Android SDK. See the License.h file:

```
constexpr int LICENSED = 0x0100;
constexpr int NOT_LICENSED = 0x0231;
constexpr int RETRY = 0x0123;
constexpr int ERROR_INVALID_PACKAGE_NAME = 1;
constexpr int ERROR_NON_MATCHING_UID = 2;
constexpr int ERROR_NOT_MARKET_MANAGED = 3;
constexpr int ERROR_CHECK_IN_PROGRESS = 4;
constexpr int ERROR_INVALID_PUBLIC_KEY = 5;
constexpr int ERROR_MISSING_PERMISSION = 6;
```

Then, Callbacks.h declares callbacks invoked from the license-checker:

```
void OnStart();
void OnLicensed( int Reason );
void OnLicenseError( int ErrorCode );
```

The main source file contains the implementations of those callbacks:

```
#include <stdlib.h>
#include "Callbacks.h"
#include "License.h"
#include "Log.h"
void OnStart()
{
  LOGI( "Hello Android NDK!" );
}
void OnLicensed( int Reason )
{
  LOGI( "OnLicensed: %i", Reason );
```

Here, terminate the application only if we are truly unlicensed:

```
  if ( Reason == NOT_LICENSED )
  {
    exit( 255 );
  }
}
void OnLicenseError( int ErrorCode )
{
  LOGI( "ApplicationError: %i", ErrorCode );
}
```

Let's dive deeper into JNI and Java code to see how these callbacks are invoked. The `LicenseChecker.cpp` file contains a one-to-one mapping of static Java methods to the previously mentioned callbacks:

```
extern "C"
{
  JNIEXPORT void JNICALL
  Java_com_packtpub_ndkmastering_AppActivity_Allow(
    JNIEnv* env, jobject obj, int Reason )
  {
    OnLicensed( Reason );
  }
  JNIEXPORT void JNICALL
  Java_com_packtpub_ndkmastering_AppActivity_DontAllow(
    JNIEnv* env, jobject obj, int Reason )
  {
    OnLicensed( Reason );
  }
  JNIEXPORT void JNICALL
  Java_com_packtpub_ndkmastering_AppActivity_ApplicationError(
    JNIEnv* env, jobject obj, int ErrorCode )
  {
    OnLicenseError( ErrorCode );
  }
}
```

We follow the code into the `AppActivity.java` file, which declares `CheckLicense()`:

```
public void CheckLicense( String BASE64_PUBLIC_KEY,
  byte[] SALT )
{
  String deviceId = Secure.getString(
    getContentResolver(), Secure.ANDROID_ID );
```

Construct the `LicenseCheckerCallback` object. The Google licensing library invokes it once done:

```
m_LicenseCheckerCallback = new AppLicenseChecker();
```

Construct `LicenseChecker` with `Policy`:

```
m_Checker = new LicenseChecker( this,
  new ServerManagedPolicy(this,
    new AESObfuscator( SALT,
```

```
        getPackageName(), deviceId) ),
      BASE64_PUBLIC_KEY);
    m_Checker.checkAccess( m_LicenseCheckerCallback );
  }
```

The Java side of the callbacks is right here, at the bottom of the class declaration:

```
public static native void Allow( int reason );
public static native void DontAllow( int reason );
public static native void ApplicationError( int errorCode );
```

The `AppLicenseChecker` class just calls these static methods to route the events to JNI code. How simple! Now, you can handle (and test) the reaction to license checking events in C++ code in a portable way. Build the sample for Android with the following commands and see for yourself:

```
>ndk-build
```

```
>ant debug
```

The runtime logs can be accessed via `logcat`. The desktop version is buildable via the `make all` command as all the samples in this book.

Flurry analytics

Let's touch one more Java-related thing and its binding to the native C++ code. Flurry.com is a popular in-app analytics service. Determination of the most used features in your app is accomplished by sending information to Flurry.com and later accessing gathered statistics on their web pages.

For example, you have several options in your application such as the different game modes: campaign, single level, or online. User selects one of the modes and an event is generated and sent to Flurry.com. We want to send those events from our C++ code.

Check out the sample application in the `5_Flurry` folder. The `main.cpp` file contains a typical usage example:

```
void OnStart()
{
  TrackEvent( "FlurryTestEvent" );
}
```

The definition of `TrackEvent()` and the difference between Android and desktop implementations is located in the `Callbacks.cpp` file:

```
extern "C"
{
  void Android_TrackEvent( const char* EventID );
};
void TrackEvent( const char* EventID )
{
  #if defined(ANDROID)
    Android_TrackEvent( EventID );
  #else
    printf( "TrackEvent: %s\n", EventID );
  #endif
}
```

Android implementation requires some JNI code to work. Have a look at the following `jni/JNI.c` file:

```
void Android_TrackEvent( const char* EventID )
{
  JAVA_ENTER();
  jstring jstr = (*env)->NewStringUTF( env, EventID );
  FindJavaStaticMethod( env, &Class, &Method,
    "com/packtpub/ndkmastering/AppActivity",
    "Callback_TrackEvent", "(Ljava/lang/String;)V" );
  (*env)->CallStaticVoidMethod( env, Class, Method, jstr );
  JAVA_LEAVE();
}
```

And `Callback_TrackEvent()` is defined in the main activity:

```
public static void Callback_TrackEvent( String EventID )
{
  if ( m_Activity == null ) return;
  m_Activity.TrackEvent( EventID );
}
public void TrackEvent( String EventID )
{
  FlurryAgent.logEvent( EventID );
}
```

Other parts of Flurry analytics API can be routed from C++ to Java and back in the similar way. We advice that you register an account on Flurry, obtain the application keys, and try running the sample yourself. Only the application keys for `FlurryAgent.init()` and `FlurryAgent.onStartSession()` should be replaced to run the application on Android. Building is straightforward, just use `ndk-build` and `ant debug`.

Summary

In this chapter, we learned how to implement lean and portable multithreading primitives, such as a reference-countered intrusive smart pointer, a worker thread, and a message pump, and use them to create simple and portable network-accessing framework in C++. A little bit of Java was touched to show how to deal with licensing and usage analytics in the native code. In the next chapter, we will step aside from networking and learn how to deal with heterogeneous filesystems using a virtual filesystem abstraction.

4
Organizing a Virtual Filesystem

In this chapter, we will implement low level abstractions to deal with OS-independent access to files and filesystems. We will show how to implement portable and transparent access to Android assets packed inside .apk files without falling back on any built-in APIs. This approach is necessary when building multi-platform applications debuggable in a desktop environment.

Mount points

The concept of mount points can be found in almost every modern filesystem. For a multi-platform C++ program, it is convenient to access files across heterogeneous storage devices in a unified way. For example, on Android, each read-only data file can be stored inside the .apk package and the developer is forced to use an Android-specific asset management API. On OSX and iOS, accessing program bundles requires yet another API, on Windows an application should store everything in its folder whose physical path also varies depending on where the application was installed.

To organize file access across different platforms, we propose a shallow class hierarchy that abstracts away the differences of file management, as shown in the following figure:

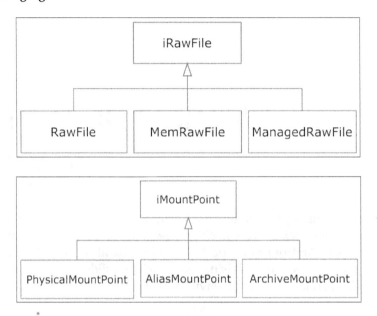

The virtual filesystem is a collection of mount points. Each mount point is an abstraction of a filesystem folder. This organization allows us to hide actual OS-specific file access routines and file name mapping from the application code. This chapter covers the description of filesystems, mount points and stream interfaces.

We define an iMountPoint interface which can resolve virtual file names and create instances of the file reading objects:

```
class iMountPoint: public iIntrusiveCounter
{
public:
```

Check whether the virtual file exists at this mount point:

```
virtual bool FileExists( const std::string& VirtualName )
   const = 0;
```

Convert a virtual file name to an absolute filename:

```
virtual std::string MapName( const std::string&
  VirtualName ) const = 0;
```

The `CreateReader()` member function creates a file reader object which implements the `iRawFile` interface introduced hereinafter in this chapter. This method is usually used only by the `clFileSystem` class:

```
virtual clPtr<iRawFile> CreateReader( const std::string&
  VirtualName ) const = 0;
```

The last two member functions get and set the internal name of this mount point. This string is used later in the `clFileSystem` interface to search and identify mount points:

```
virtual void SetName( const std::string& N ) { FName = N; }
  virtual std::string GetName() const { return FName; }
private:
  std::string FName;
};
```

Our virtual filesystem is implemented as a linear collection of mount points. The `clFileSystem::CreateReader()` method here creates an `iIStream` object which encapsulates access to file data:

```
clPtr<iIStream> CreateReader( const std::string& FileName ) const;
```

The `Mount()` method adds a physical (by *physical* we mean an OS-specific path) path to the list of mount points. If the `PhysicalPath` value represents a folder of the local filesystem, a `clPhysicalMountPoint` instance is created. If `PhysicalPath` is the name of a `.zip` or `.apk` file, the `clArchiveMountPoint` instance is added to the list of mount points. Definitions of the `clPhysicalMountPoint` and `ArchiveMountPoint` classes can be found in the example `1_ArchiveFileAccess` from the code bundle:

```
void Mount( const std::string& PhysicalPath );
```

`VirtualNameToPhysical()` converts our virtual path to an OS-specific system file path:

```
std::string VirtualNameToPhysical(
  const std::string& Path ) const;
```

The `FileExists()` method inspects each mount point to see if the file exists in one of them:

```
bool FileExists( const std::string& Name ) const;
```

The private part of the `clFileSystem` class is responsible for management of the internal list of mount points. The `FindMountPoint()` method searches the mount point containing a file named `FileName`. The `FindMountPointByName()` method is used internally to allow aliasing of file names. `AddMountPoint()` checks if the supplied mount point is unique and, if it is, adds it to the `FMountPoints` container:

```
private:
  clPtr<iMountPoint> FindMountPointByName(
    const std::string& ThePath );
  void AddMountPoint( const clPtr<iMountPoint>& MP );
  clPtr<iMountPoint> FindMountPoint(
    const std::string& FileName ) const;
```

Finally, the collection of mount points is stored in `std::vector`:

```
  std::vector< clPtr<iMountPoint> > FMountPoints;
};
```

When we want to access a file in our application code, we do it via the filesystem object `g_FS`:

```
auto f = g_FS->CreateReader( "test.txt" );
```

Mount points and streams

On Android, the `test.txt` file is most likely to reside in the `.apk` package and a lot of work needs to happen within the `CreateReader()` call. The data for `test.txt` is extracted and an instance of `clMemFileMapper` is created. Let's travel down the hidden pipeline of file operations.

The code for `CreateReader()` is simple. First, we convert the slashes and backslashes in the path to match those of the underlying operating system. Then a mount point is found which hosts the file named `FileName`. Finally, an instance of `clFileMapper` is created. This class implements the `iIStream` interface. Let's take a closer look at all these classes:

```
clPtr<iIStream> clFileSystem::CreateReader(
  const std::string& FileName ) const
{
  std::string Name = Arch_FixFileName( FileName );
  clPtr<iMountPoint> MountPoint = FindMountPoint( Name );
```

Here we use the Null Object pattern (`http://en.wikipedia.org/wiki/Null_Object_pattern`) to define neutral behavior in the case of a non-existent file. The `clNullRawFile` class represents an empty file not tied to any real device:

```
    if ( !MountPoint ) { return make_intrusive<clFileMapper>(
      make_intrusive<clNullRawFile>() ); }
    return make_intrusive<clFileMapper>( MountPoint->CreateReader(
      Name ) );
}
```

The `FindMountPoint()` method iterates the collection of mount points to find the one which contains a file with a given name:

```
clPtr<iMountPoint> clFileSystem::FindMountPoint( const
  std::string& FileName ) const
{
  if ( FMountPoints.empty() )
  {
    return nullptr;
  }
  if ( ( *FMountPoints.begin() )->FileExists( FileName ) )
  {
    return ( *FMountPoints.begin() );
  }
```

Iterate mount points in reverse order, so that the most recently mounted paths will be checked first:

```
    for ( auto i = FMountPoints.rbegin();
      i != FMountPoints.rend(); ++i )
    {
      if ( ( *i )->FileExists( FileName ) )
      {
        return ( *i );
      }
    }
    return *( FMountPoints.begin() );
}
```

The `clFileSystem` class delegates most of its work to individual `iMountPoint` instances. For example, the check for file existence is performed by finding the appropriate `iMountPoint` object and asking it if a file exists at that point:

```
bool clFileSystem::FileExists( const std::string& Name ) const
{
  if ( Name.empty() || Name == "." ) { return false; }
  clPtr<iMountPoint> MP = FindMountPoint( Name );
  return MP ? MPD->FileExists( Name ) : false;
}
```

The physical file name is also found using the appropriate `iMountPoint` instance:

```
std::string clFileSystem::VirtualNameToPhysical(
  const std::string& Path ) const
{
  if ( FS_IsFullPath( Path ) ) { return Path; }
  clPtr<iMountPoint> MP = FindMountPoint( Path );
  return ( !MP ) ? Path : MP->MapName( Path );
}
```

Physical file names are not used directly to access files. For example, if an archive was mounted and we wanted to access the file in the archive, the physical path of that file is meaningless to the operating system. Instead, everything is abstracted by mount points and physical file names are only used as identifiers in our applications.

The new mount point is added to the collection only if it is unique; there is no reason to allow duplicates.

```
void clFileSystem::AddMountPoint( const clPtr<iMountPoint>& MP )
{
  if ( !MP ) { return; }
  if ( std::find( FMountPoints.begin(),
    FMountPoints.end(), MP ) == FMountPoints.end() )
  {
    FMountPoints.push_back( MP );
  }
}
```

The code for `clFileSystem::Mount()` selects which mount point type to instantiate:

```
void clFileSystem::Mount( const std::string& PhysicalPath )
{
  clPtr<iMountPoint> MP;
```

We use a simple hardcoded logic here. If an path ends with a `.zip` or `.apk` substring, we instantiate `clArchiveMountPoint`:

```
  if ( Str::EndsWith( PhysicalPath, ".apk" ) ||
    Str::EndsWith( PhysicalPath, ".zip" ) )
  {
    auto Reader = make_intrusive<clArchiveReader>();
    bool Result = Reader->OpenArchive(
      CreateReader( PhysicalPath ) );
    MP = make_intrusive<clArchiveMountPoint>( Reader );
  }
  else
```

Otherwise, we check if `clPhysicalPath` exists and then create
`clPhysicalMountPoint`:

```
{
  #if !defined( OS_ANDROID )
    if ( !FS_FileExistsPhys( PhysicalPath ) )
    return;
  #endif
    MP = make_intrusive<clPhysicalMountPoint>(PhysicalPath );
}
```

If the mount point creation succeeds, we set its name and add it to the collection:

```
  MP->SetName( PhysicalPath );
  AddMountPoint( MP );
}
```

We will return to the mount point implementations later. Right now, we turn to the
streams. Actual read access to files is done through the `iIStream` interface:

```
class iIStream: public iIntrusiveCounter
{
public:
```

The following two methods get virtual and physical file names respectively:

```
virtual std::string GetVirtualFileName() const = 0;
virtual std::string GetFileName() const = 0;
```

The `Seek()` method sets the absolute reading position; `GetSize()` and `GetPos()`
determine the size and the current reading position while `Eof()` checks if the end of
the file has been reached:

```
virtual void   Seek( const uint64 Position ) = 0;
virtual uint64 GetSize() const = 0;
virtual uint64 GetPos() const = 0;
virtual bool   Eof() const = 0;
```

The `Read()` method reads the block of data with the specified `Size` into the untyped
memory buffer `Buf`:

```
virtual uint64 Read( void* Buf, const uint64 Size ) = 0;
```

The last two methods use memory mapping for array-like access to the file data. The first one returns a pointer to the shared memory corresponding to this file:

```
virtual const ubyte* MapStream() const = 0;
```

The second one returns a pointer to the memory starting from the current file position. This is convenient for seamless switching between block and memory-mapped styles of access:

```
virtual const ubyte* MapStreamFromCurrentPos() const = 0;
};
```

To avoid UI thread blocking, these methods should be usually called on the worker threads.

All work to access the physical files is done within the `clFileMapper` class. It is an implementation of the `iIStream` interface which delegates all the I/O to an object implementing the `iRawFile` interface. `iRawFile` itself is not used directly in the application code, so let's look at the `clFileMapper` class first:

```
class clFileMapper: public iIStream
{
public:
```

The constructor just stores the reference to the `iRawFile` instance and resets the read pointer:

```
explicit FileMapper( clPtr<iRawFile> File ):
   FFile( File ), FPosition( 0 ) {}
virtual ~FileMapper() {}
```

The `GetVirtualFileName()` and `GetFileName()` methods use the instance of `iRawFile` to get virtual and physical file names respectively:

```
virtual std::string GetVirtualFileName() const
{ return FFile->GetVirtualFileName(); }
virtual std::string GetFileName() const
{ return FFile->GetFileName(); }
```

The `Read()` method emulates the `std::ifstream.read` and the `read()` routines from `libc`. It might seem unusual, but reading is done with the `memcpy` call which accesses the memory-mapped file. The description of `iRawFile::GetFileData()` will clarify things:

```
virtual uint64 Read( void* Buf, uint64 Size )
{
```

```
    uint64 RealSize = ( Size > GetBytesLeft() ) ?
      GetBytesLeft() : Size;
    if ( !RealSize ) { return 0; }
    memcpy( Buf, ( FFile->GetFileData() + FPosition ),
      static_cast<size_t>( RealSize ) );
    FPosition += RealSize;
    return RealSize;
  }
```

Positioning and memory mapping are all delegated to the underlying
iRawFile instance:

```
    virtual void Seek( const uint64 Position)
    { FPosition = Position; }
    virtual uint64 GetSize() const
    { return FFile->GetFileSize(); }
    virtual bool Eof() const
    { return ( FPosition >= FFile->GetFileSize() ); }
    virtual const ubyte* MapStream() const
    { return FFile->GetFileData(); }
    virtual const ubyte* MapStreamFromCurrentPos() const
    { return ( FFile->GetFileData() + FPosition ); }
```

The private section contains a reference to iRawFile and the current
reading position:

```
private:
  clPtr<iRawFile> FFile;
  uint64 FPosition;
};
```

Now we can declare the iRawFile interface, which is very simple:

```
class iRawFile: public iIntrusiveCounter
{
public:
  iRawFile() {}
  virtual ~iRawFile() {}
```

The first four methods get and set the virtual and physical file names:

```
    std::string GetVirtualFileName() const
    { return FVirtualFileName; }
    std::string  GetFileName() const
    { return FFileName; }
```

```
void SetVirtualFileName( const std::string& VFName )
{ FVirtualFileName = VFName; }
void SetFileName( const std::string& FName )
{ FFileName = FName; }
```

The essence of this interface is in the following two methods, which get the raw pointer to the file data and the size of the file:

```
virtual const ubyte* GetFileData() const = 0;
virtual uint64 GetFileSize() const = 0;
```

The private section contains strings with file names:

```
private:
  std::string    FFileName;
  std::string    FVirtualFileName;
};
```

Having declared all the interfaces, we may proceed to their implementation.

Accessing files on the host filesystems

We start from the clRawFile class, which uses OS-specific memory-mapping routines to map files into the memory:

```
class clRawFile: public iRawFile
{
public:
  RawFile() {}
  virtual ~RawFile() { Close(); }
```

The Open() member function does most of the heavy lifting. It stores physical and virtual file names, opens a file handle and creates a mapped view of the file:

```
bool Open( const std::string& FileName,
  const std::string& VirtualFileName )
{
  SetFileName( FileName );
  SetVirtualFileName( VirtualFileName );
  FSize = 0;
  FFileData = nullptr;
```

With Windows, we use `CreateFileA()` to open the file. As usual, we enclose the OS-specific parts in `#ifdef` blocks.:

```
#ifdef _WIN32
  FMapFile = CreateFileA( FFileName.c_str(), GENERIC_READ,
    FILE_SHARE_READ, nullptr, OPEN_EXISTING,
    FILE_ATTRIBUTE_NORMAL | FILE_FLAG_RANDOM_ACCESS,
    nullptr );
```

Once the file is opened, we create a mapping object and retrieve a pointer to file data using the `MapViewOfFile()` system call:

```
FMapHandle = CreateFileMapping( FMapFile,
  nullptr, PAGE_READONLY, 0, 0, nullptr );
FFileData = ( ubyte* )MapViewOfFile( FMapHandle,
  FILE_MAP_READ, 0, 0, 0 );
```

If something goes wrong, close the handle and cancel the operation:

```
if ( !FFileData )
{
  CloseHandle( ( HANDLE )FMapHandle );
  return false;
}
```

To prevent reading past the end of the file, we should retrieve the size of the file. This is how it's done with Windows:

```
DWORD dwSizeLow = 0, dwSizeHigh = 0;
dwSizeLow = ::GetFileSize( FMapFile, &dwSizeHigh );
FSize = ( ( uint64 )dwSizeHigh << 32 )
  | ( uint64 )dwSizeLow;
```

With Android, we use `open()` to initialize the file handle and `fstat()` to get its size:

```
#else
  FFileHandle = open( FileName.c_str(), O_RDONLY );
  struct stat FileInfo;
```

If `fstat()` succeeds, we can retrieve its size. If the file has non-zero size, we call the `mmap()` function to map the file into memory:

```
if ( !fstat( FFileHandle, &FileInfo ) )
{
  FSize = static_cast<uint64_t>( FileInfo.st_size );
```

Make sure we do not call `mmap()` for zero-sized files:

```
if ( FSize )
  FFileData = ( uint8_t* )( mmap( nullptr, FSize,
    PROT_READ, MAP_PRIVATE, FFileHandle, 0 ) );
}
```

We can immediately close the file handle once we have the `mmap`-ed memory block. This is the standard way:

```
    close( FFileHandle );
  #endif
    return true;
}
```

The `Close()` method unmaps the memory block and closes the file handle:

```
void Close()
{
```

With Windows, we use the `UnmapViewOfFile()` and `CloseHandle()` system calls:

```
#ifdef _WIN32
  if ( FFileData  ) { UnmapViewOfFile( FFileData ); }
  if ( FMapHandle ) { CloseHandle( (HANDLE)FMapHandle ); }
  CloseHandle( ( HANDLE )FMapFile );
```

With Android, we call the `munmap()` function:

```
#else
  if ( FFileData )
  {
    munmap( reinterpret_cast<void*>( FFileData ), FSize );
  }
  #endif
}
```

The rest of the `clRawFile` class contains two easy methods that return the file data pointer and file size. The private part declares file handles, file size and the data pointer:

```
  virtual const ubyte* GetFileData() const { return FFileData; }
  virtual uint64       GetFileSize() const { return FSize; }
private:
  #ifdef _WIN32
    HANDLE    FMapFile;
```

```
   HANDLE      FMapHandle;
#else
   int         FFileHandle;
#endif
  ubyte*       FFileData;
  uint64       FSize;
};
```

To access physical folders in our virtual filesystem using the clFileSystem class, we declare the clPhysicalMountPoint class representing a single folder on the host filesystem:

```
class clPhysicalMountPoint: public iMountPoint
{
public:
```

The constructor of clPhysicalMountPoint fixes the physical folder path by adding a trailing path separator which is the slash or the backslash character depending on the conventions of the underlying OS:

```
clPhysicalMountPoint( const std::string& PhysicalName ):
  FPhysicalName( PhysicalName )
{
  Str_AddTrailingChar( &FPhysicalName, PATH_SEPARATOR );
}
virtual ~PhysicalMountPoint() {}
```

The FileExists() method uses an OS-dependent routine to check if the file exists:

```
virtual bool FileExists( const std::string& VirtualName )
  const override
{
  return FS_FileExistsPhys( MapName( VirtualName ) );
}
```

MapName() converts the virtual file into the physical file name by adding the FPhysicalName prefix. The FS_IsFullPath() routine is defined in the following code:

```
virtual std::string  MapName( const std::string& VirtualName )
  const override
{
  return FS_IsFullPath( VirtualName ) ?
    VirtualName : ( FPhysicalName + VirtualName );
}
```

Instances of `clRawFile` are created in the `clPhysicalMountPoint::CreateReader()` method:

```
virtual clPtr<iRawFile> CreateReader(
   const std::string& VirtualName ) const override
{
   std::string PhysName = MapName( VirtualName );
   auto File = make_intrusive<clRawFile>();
   if ( File->Open( FS_ValidatePath( PhysName ), VirtualName ) )
   { return File; }
   return make_intrusive<clNullRawFile>();
}
```

The private part of the class contains a physical name of the folder:

```
private:
   std::string FPhysicalName;
};
```

To complete this code, we have to implement some service routines. The first one is `FS_IsFullPath()`, which checks if the path is an absolute one. For Android, this means the path starts from the / character and, for Windows, the full path must start with the <drive>:\ substring, where <drive> is the drive letter:

```
inline bool FS_IsFullPath( const std::string& Path )
{
   return ( Path.find( ":\\" ) != std::string::npos ||
   #if !defined( _WIN32 )
      ( Path.length() && Path[0] == '/' ) ||
   #endif
      Path.find( ":/" )  != std::string::npos ||
      Path.find( ".\\" ) != std::string::npos );
}
```

`FS_ValidatePath()` replaces each slash or backslash character with the platform-specific PATH_SEPARATOR:

```
inline std::string FS_ValidatePath( const std::string& PathName )
{
   std::string Result = PathName;
   for ( size_t i = 0; i != Result.length(); ++i )
      if ( Result[i] == '/' || Result[i] == '\\' )
      {
```

```
        Result[i] = PATH_SEPARATOR;
      }
    return Result;
  }
```

To check if the file exists, we use the `stat()` routine whose syntax differs slightly with Windows and Android:

```
inline bool FS_FileExistsPhys( const std::string& PhysicalName )
{
  #ifdef _WIN32
    struct _stat buf;
    int Result = _stat( FS_ValidatePath( PhysicalName ).c_str(),
      &buf );
  #else
    struct stat buf;
    int Result = stat( FS_ValidatePath( PhysicalName ).c_str(),
      &buf );
  #endif
    return Result == 0;
}
```

`PATH_SEPARATOR` is a platform-specific character constant:

```
#if defined( _WIN32 )
  const char PATH_SEPARATOR = '\\';
#else
  const char PATH_SEPARATOR = '/';
#endif
```

This code is enough to access the files stored directly on the host filesystem. Let us proceed with other abstractions to get to Android `.apk` packages.

In-memory files

The following implementation of the `iRawFile` interface encapsulates access to untyped memory blocks as file access. We will use this class to access uncompressed data in archives.

```
class clMemRawFile: public iRawFile
{
public:
```

The parameterized constructor initializes the pointer to a data buffer and its size:

```
clMemRawFile(
   const uint8_t* BufPtr, size_t BufSize, bool OwnsBuffer )
: FOwnsBuffer( OwnsBuffer )
, FBuffer( BufPtr )
, FBufferSize( BufSize )
{}
```

The memory mapping is trivial for a memory block, we just return the stored raw pointer:

```
virtual const uint8_t* GetFileData() const override
{ return FBuffer; }
virtual uint64_t GetFileSize() const override
{ return FBufferSize; }
private:
   const uint8_t* FBuffer;
   size_t FBufferSize;
};
```

We will return to this class once we deal with archive file reading.
Now, let's get familiar with one more important concept required to access
.apk packages transparently.

Aliasing

The file abstractions mentioned in the preceding section are very powerful. They can be used to create nested mount points to access files packed within other files. Let's demonstrate the flexibility of this approach by defining clAliasMountPoint, which acts like a symbolic link on Unix or NTFS filesystems.

The implementation redirects each iMountPoint:: method call to another mount point instance while transforming the file name on the fly by prepending each virtual file name we want to access with a specified FAlias prefix:

```
class clAliasMountPoint: public iMountPoint
{
public:
   explicit clAliasMountPoint( const clPtr<iMountPoint>& Src )
   : Falias(), FMP( Src )
   {}
```

```
    virtual bool FileExists( const std::string& VirtualName )
      const { return FMP->FileExists( FAlias + VirtualName ); }
    virtual std::string MapName( const std::string& VirtualName )
      const { return FMP->MapName( FAlias + VirtualName ); }
    virtual clPtr<iRawFile> CreateReader( const std::string& VirtualName )
      const { return FMP->CreateReader( FAlias + VirtualName ); }
  private:
    std::string FAlias;
    clPtr<iMountPoint> FMP;
  };
```

We add the `FileSystem::AddAlias()` member function which decorates an existing mount point's file names by concatenating them with the `FAlias` prefix:

```
void clFileSystem::AddAlias( const std::string& SrcPath,
  const std::string& Alias )
{
  if (clPtr<iMountPoint> MP = FindMountPointByName( SrcPath ) )
    AddMountPoint(new AliasMountPoint( MP, Alias ) );
}
```

This mechanism can be used to transparently remap paths such as `assets/` to the root of our filesystem, which is essential to the functionality of our applications on Android.

Writing files

Before proceeding to the more complex stuff of archive unpacking, let's take a short break and take a look at how to write to a file. We use the `iOStream` interface which declares only four pure virtual methods. The `GetFileName()` method returns the virtual file name. The `Seek()` method sets the writing position and `GetFilePos()` returns it. The `Write()` method takes an untyped memory buffer and writes it to the output stream:

```
class iOStream: public iIntrusiveCounter
{
public:
  iOStream() {};
  virtual ~iOStream() {};
  virtual std::string GetFileName() const = 0;
  virtual void    Seek( const uint64 Position ) = 0;
  virtual uint64 GetFilePos() const = 0;
  virtual uint64 Write(const void* Buf, const uint64 Size) = 0;
};
```

The only implementation of `iOStream` we provide here is `clMemFileWriter`, which treats an untyped memory block as an output stream. This class is used to access data in `.zip` files. First, the data is unpacked, then it is wrapped using `clMemRawFile`:

```
class clMemFileWriter: public iOStream
{
public:
```

The actual underlying memory block is RAII-managed (https://en.wikipedia. org/wiki/Resource_Acquisition_Is_Initialization) by the `clBlob` object stored inside this class:

```
clMemFileWriter()
: FBlob( make_intrusive<clBlob>() )
, FFileName()
, FPosition( 0 )
{}
explicit clMemFileWriter( const clPtr<clBlob>& Blob )
: FBlob( Blob )
, FFileName()
, FPosition( 0 )
{}
```

The `Seek()` method increments the current writing position:

```
virtual void Seek( const uint64 Position )
{
    FPosition = ( Position > FBlob->GetSize() ) ?
      FBlob->GetSize() - 1 : Position;
}
```

The `Write()` method redirects to the `clBlob` object:

```
virtual uint64_t Write( const void* Buf, uint64_t Size )
  override
{
    return FBlob->AppendBytes( Buf, static_cast<size_t>( Size ) );
}
```

The accompanying source code contains the implementation of the `clFileWriter` class which contains the `Open()` method similar to `clRawFile::Open()`. The `Write()` method uses system I/O routines to write data to a physical file.

Now we have enough scaffolding code to proceed further with `.zip` archives.

Accessing the archive files

Since .apk is just a fancy .zip archive, we use the ZLib library by Jean-loup Gailly combined with the MiniZIP library to retrieve compressed files from it. The complete source code is about 500 kilobytes in size so we provide two files, libcompress.c and libcompress.h, which are easily integrated into any build process. Our goal is to implement the clArchiveMountPoint which enumerates files in an archive, decompresses the data for the specific file, and creates a clMemFileMapper to read its data. To do this, we need to introduce a helper class, clArchiveReader, which reads and decompresses .zip archives:

```
class clArchiveReader: public iIntrusiveCounter
{
private:
```

The private sFileInfo structure is defined in the clArchiveReader class and encapsulates a pack of useful file properties together with the pointer to compressed file data:

```
struct sFileInfo
{
  /// offset to the file
  uint64 FOffset;
  /// uncompressed file size
  uint64 FSize;
  /// compressed file size
  uint64 FCompressedSize;
  /// Compressed data
  void* FSourceData;
};
```

The private section of the clArchiveReader class contains a collection of sFileInfo structures in a FFileInfos field, a vector of uppercased file names FFileNames, a vector of in-archive file names FReadFileNames, and an std::map object, which maps each file name to an index in the extracted files vector FExtractedFromArchive:

```
std::vector<sFileInfo> FFileInfos;
std::vector<std::string> FFileNames;
std::vector<std::string> FRealFileNames;
mutable std::map<std::string, int> FFileInfoIdx;
std::map<int, const void*> FExtractedFromArchive;
```

The FSourceFile field holds a pointer to the source file stream of the .apk file:

```
    clPtr<iIStream> FSourceFile;
public:
  clArchiveReader()
  : FFileInfos()
  , FRealFileNames()
  , FFileInfoIdx()
  , FSourceFile()
  {}
  virtual ~clArchiveReader()
  { CloseArchive(); }
```

The OpenArchive() member function invokes Enumerate_ZIP() to fill the FFileInfos container. CloseArchive() performs some required cleanup:

```
  bool OpenArchive( const clPtr<iIStream>& Source )
  {
   if ( !Source ) { return false; }
   if ( !CloseArchive() ) { return false; }
   if ( !Source->GetSize() ) { return false ; }
   FSourceFile = Source;
   return Enumerate_ZIP();
  }
  bool CloseArchive()
  {
    FFileInfos.clear();
    FFileInfoIdx.clear();
    FFileNames.clear();
    FRealFileNames.clear();
    ClearExtracted();
    FSourceFile = nullptr;
    return true;
  }
```

The long ExtractSingleFile() method is described in detail below. It accepts a name of a compressed file from the archive and an iOStream object which contains the file data. The AbortFlag pointer to an atomic Boolean flag is used for multi-threaded decompression. It is polled from time to time in the decompressor. If the value is set to true, the internal decompression loop terminates prematurely and ExtractSingleFile() returns false.

The `Progress` pointer is used to update the progress of the decompression process, which should be atomic as well. An optional `Password` parameter can be provided, if an archive is encrypted:

```
bool ExtractSingleFile( const std::string& FileName,
   const std::string& Password, std::atomic<int>* AbortFlag,
   std::atomic<float>* Progress, const clPtr<iOStream>& Out );
```

The next two methods use the `FFileInfos` vector to check if a file exists within this archive and retrieve its decompressed size:

```
bool FileExists( const std::string& FileName ) const
{
   return GetFileIdx( FileName ) > -1;
}
uint64 GetFileSizeIdx( const std::string& FileName ) const
{
   return ( Idx > -1 ) ? FFileInfos[ Idx ].FSize : 0;
}
```

The `GetFileDataIdx()` method first checks if the file was already decompressed. In this case, the pointer from `FExtractedFromArchive` is returned:

```
const void* GetFileDataIdx( int Idx )
{
   if ( Idx <= -1 ) { return nullptr; }
   if ( FExtractedFromArchive.count( Idx ) > 0 )
   {
      return FExtractedFromArchive[Idx]->GetDataConst();
   }
```

If the file was not decompressed yet, the `GetFileData_ZIP()` function is called and an unpacked memory block from `clBlob` is returned:

```
   auto Blob = GetFileData_ZIP( Idx );
   if ( Blob )
   {
      FExtractedFromArchive[Idx] = Blob;
      return Blob->GetDataConst();
   }
   return nullptr;
}
```

The GetFileIdx() method maps a FileName to an internal index into the FFileInfos vector. It uses the auxiliary FFileInfoIdx object to store the string-to-index correspondences:

```
int GetFileIdx( const std::string& FileName ) const
{
  return ( FFileInfoIdx.count( FileName ) > 0 ) ?
    FFileInfoIdx[ FileName ] : -1;
}
```

The last two public functions return the number of files in the archive and the name of each file:

```
size_t GetNumFiles() const { return FFileInfos.size(); }
std::string GetFileName( int Idx ) const
{ return FFileNames[Idx]; }
```

The private section of the clArchiveReader class declares internal methods for decompressed data management. The Enumerate_ZIP() method fills the FFileInfos container by reading the archive header. The GetFileData_ZIP() member function extracts file data from the archive:

```
private:
  bool Enumerate_ZIP();
  const void* GetFileData_ZIP( size_t Idx );
```

The ClearExtracted() method is invoked from CloseArchive(). It frees the allocated memory for each extracted file. Everything here is RAII-managed using the clBlob class:

```
void ClearExtracted()
{
  FExtractedFromArchive.clear();
}
```

Let's look into the GetFileData_ZIP() method implementation which uses the ExtractSingleFile() method:

```
clPtr<clBlob> clArchiveReader::GetFileData_ZIP( int Idx )
{
  if ( FExtractedFromArchive.count( Idx ) > 0 )
  {
    return FExtractedFromArchive[ Idx ];
  }
```

The `clMemFileWriter` object is created, which contains the decompressed data:

```
clPtr<clMemFileWriter> Out =
  clFileSystem::CreateMemWriter( "mem_blob",
    FFileInfos[ Idx ].FSize );
```

`ExtractSingleFile()` handles decompression. Here we use a blocking call (the `AbortFlag` parameter is `nullptr`) and an empty password:

```
if ( ExtractSingleFile( FRealFileNames[ Idx ], "",
  nullptr, nullptr, Out ) )
{
```

If the call succeeds, we return the decompressed contents from the `clMemFileWriter` object:

```
  return Out->GetBlob();
}
return make_intrusive<clBlob>();
}
```

The `ExtractSingleFile()` method creates the `zlib` reader object, positions the reader at the beginning of the compressed file data and calls the `ExtractCurrentFile_ZIP()` method to perform the actual decompression:

```
bool clArchiveReader::ExtractSingleFile(
  const std::string& FileName, const std::string& Password,
  std::atomic<int>* AbortFlag, std::atomic<float>* Progress,
  const clPtr<iOStream>& Out )
{
  std::string ZipName = FileName;
  std::replace( ZipName.begin(), ZipName.end(), '\\', '/' );
  clPtr<iIStream> TheSource = FSourceFile;
  FSourceFile->Seek( 0 );
```

We create the internal structure to allow `zlib` to read from our `iIStream` objects. The same thing is done later in `Enumerate_ZIP()`. The `fill_functions()` routine and all related callbacks are described below in this section:

```
zlib_filefunc64_def ffunc;
fill_functions( TheSource.GetInternalPtr(), &ffunc );
unzFile UnzipFile = unzOpen2_64( "", &ffunc );
if ( unzLocateFile(UnzipFile, ZipName.c_str(), 0) != UNZ_OK )
{
```

Return `false` if the file is not found within the archive:

```
    return false;
  }
```

Once we have positioned the reader, we call the `ExtractCurrentFile_ZIP()` method:

```
    int ErrorCode = ExtractCurrentFile_ZIP( UnzipFile,
      Password.empty() ? nullptr : Password.c_str(),
      AbortFlag, Progress, Out );
    unzClose( UnzipFile );
    return ErrorCode == UNZ_OK;
  }
```

The core of our decompressor lays inside `ExtractCurrentFile_Zip()`. Taking a memory block as an input, it reads decompressed bytes of the file and writes them into the output stream:

```
int ExtractCurrentFile_ZIP( unzFile UnzipFile,
  const char* Password, std::atomic<int>* AbortFlag,
  std::atomic<float>* Progress, const clPtr<iOStream>& Out )
{
  char FilenameInzip[1024];
  unz_file_info64 FileInfo;
```

The `unzGetCurrentFileInfo64()` function retrieves the uncompressed file size. We use it to count the total progress and write it into the `Progress` parameter:

```
    int ErrorCode = unzGetCurrentFileInfo64( UnzipFile,
      &FileInfo, FilenameInzip, sizeof( FilenameInzip ),
      nullptr, 0, nullptr, 0 );
    if ( ErrorCode != UNZ_OK ) { return ErrorCode; }
```

The `unzOpenCurrentFilePassword()` call initializes the decompression process:

```
    ErrorCode = unzOpenCurrentFilePassword( uf, password );
    if ( ErrorCode != UNZ_OK ) { return err; }
```

The final part of the method is a loop which reads a packet of decompressed bytes and calls the `iOStream::Write` method of the `Out` object:

```
    uint64_t FileSize = ( uint64_t )FileInfo.uncompressed_size;
```

In our example implementation based on memory-mapped files, we cast the 64-bit file size to `size_t`. This essentially breaks support of files greater than 2Gb in size on 32-bit targets. However, this tradeoff is acceptable in most real-world mobile applications, unless you are writing the universal `.zip` decompressor, of course:

```
Out->Reserve( ( size_t )FileSize );
unsigned char Buffer[ WRITEBUFFERSIZE ];
uint64_t TotalBytes = 0;
int BytesRead = 0;
do
{
```

Optionally we may break from the loop if the `AbortFlag` pointer (set from the other thread) instructs us to do so:

```
if ( AbortFlag && *AbortFlag ) break;
```

The `unzReadCurrentFile()` function performs decompression to the output stream:

```
BytesRead = unzReadCurrentFile( UnzipFile,
  Buffer, WRITEBUFFERSIZE );
if ( BytesRead < 0 ) { break; }
if ( BytesRead > 0 )
{
  TotalBytes += BytesRead;
  Out->Write( Buffer, BytesRead );
}
```

After writing the decompressed data, we update the `Progress` counter accordingly:

```
if ( Progress )
{
  *Progress = (float)TotalBytes / (float)FileSize;
}
}
while ( BytesRead > 0 );
```

At the end, we close the `UnzipFile` reader object:

```
ErrorCode = unzCloseCurrentFile( UnzipFile );
return ErrorCode;
}
```

The enumeration of files in an archive is done by yet another member function called
`Enumerate_ZIP()`:

```
bool Enumerate_ZIP()
{
  clPtr<iIStream> TheSource = FSourceFile;
  FSourceFile->Seek( 0 );
```

First, we fill the callbacks required by `zlib` to read the custom file stream, in this case
our `iIStream` objects:

```
zlib_filefunc64_def ffunc;
fill_functions( TheSource.GetInternalPtr(), &ffunc );
unzFile UnzipFile = unzOpen2_64( "", &ffunc );
```

Then, the header of the archive is read in order to determine the number of
compressed files:

```
unz_global_info64 gi;
int ErrorCode = unzGetGlobalInfo64( uf, &gi );
```

For each compressed file, we extract the information which we reuse later for
decompression:

```
for ( uLong i = 0; i < gi.number_entry; i++ )
{
  if ( ErrorCode != UNZ_OK ) { break; }
  char filename_inzip[256];
  unz_file_info64 file_info;
  ErrorCode = unzGetCurrentFileInfo64( UnzipFile,
    &file_info, filename_inzip,
    sizeof(filename_inzip), nullptr, 0, nullptr, 0 );
  if ( ErrorCode != UNZ_OK ) { break; }
  if ( ( i + 1 ) < gi.number_entry )
  {
    ErrorCode = unzGoToNextFile( UnzipFile );
    if ( ErrorCode != UNZ_OK ) { break; }
  }
```

In each iteration, we fill the `sFileInfo` structure and store it in the
`FFileInfos` vector:

```
sFileInfo Info;
Info.FOffset = 0;
Info.FCompressedSize = file_info.compressed_size;
Info.FSize = file_info.uncompressed_size;
FFileInfos.push_back( Info );
```

All the backslashes in the file name are converted into the characters that separate elements of the path within the archive. The `FFileInfoIdx` map is filled for a quick lookup of the file index:

```
        std::string TheName = Arch_FixFileName(filename_inzip);
        FFileInfoIdx[ TheName ] = ( int )FFileNames.size();
        FFileNames.emplace_back( TheName );
        FRealFileNames.emplace_back( filename_inzip );
    }
```

Finally, we clean up the `zlib` reader object and return the success code:

```
        unzClose( UnzipFile );
        return true;
    }
```

Let's take a closer look at the `fill_functions()` method. The memory block is contained in `iIStream`, so we implement a number of callbacks required by `zlib` to work with our stream class. The first method `zip_fopen()` does the preparation of `iIStream`:

```
    static voidpf ZCALLBACK zip_fopen ( voidpf opaque, const void*
      filename, int mode )
    {
      ( ( iIStream* )opaque )->Seek( 0 );
      return opaque;
    }
```

The reading of bytes from `iIStream` is implemented in `zip_fread()`:

```
    static uLong ZCALLBACK zip_fread ( voidpf opaque, voidpf stream,
      void* buf, uLong size )
    {
      iIStream* S = ( iIStream* )stream;
      int64 CanRead = ( int64 )size;
      int64 Sz = S->GetSize();
      int64 Ps = S->GetPos();
      if ( CanRead + Ps >= Sz ) { CanRead = Sz - Ps; }
      if ( CanRead > 0 ) { S->Read( buf, ( uint64 )CanRead ); }
      else { CanRead = 0; }
      return ( uLong )CanRead;
    }
```

The zip_ftell() function tells the current position in iIStream:

```
static ZPOS64_T ZCALLBACK zip_ftell(voidpf opaque, voidpf
  stream)
{
  return ( ZPOS64_T )( ( iIStream* )stream )->GetPos();
}
```

The zip_fseek() routine sets the reading pointer, just like libc's fseek():

```
static long ZCALLBACK zip_fseek ( voidpf  opaque, voidpf stream,
  ZPOS64_T offset, int origin )
{
  iIStream* S = ( iIStream* )stream;
  int64 NewPos = ( int64 )offset;
  int64 Sz = ( int64 )S->GetSize();
  switch ( origin )
  {
    case ZLIB_FILEFUNC_SEEK_CUR:
      NewPos += ( int64 )S->GetPos(); break;
    case ZLIB_FILEFUNC_SEEK_END:
      NewPos = Sz - 1 - NewPos; break;
    case ZLIB_FILEFUNC_SEEK_SET: break;
    default:  return -1;
  }
  if ( NewPos >= 0 && ( NewPos < Sz ) )
  {
    S->Seek( ( uint64 )NewPos );
  }
  else
  {
    return -1;
  }
  return 0;
}
```

For the iIstream class, fclose(), and ferror(), analogues are trivial:

```
static int ZCALLBACK zip_fclose( voidpf opaque, voidpf stream )
{
  return 0;
}
static int ZCALLBACK zip_ferror( voidpf opaque, voidpf stream )
{
  return 0;
}
```

A helper `fill_functions()` routine fills the callback structure used by `zlib`:

```
void fill_functions( iIStream* Stream, zlib_filefunc64_def*
  pzlib_filefunc_def )
{
  pzlib_filefunc_def->zopen64_file = zip_fopen;
  pzlib_filefunc_def->zread_file = zip_fread;
  pzlib_filefunc_def->zwrite_file = NULL;
  pzlib_filefunc_def->ztell64_file = zip_ftell;
  pzlib_filefunc_def->zseek64_file = zip_fseek;
  pzlib_filefunc_def->zclose_file = zip_fclose;
  pzlib_filefunc_def->zerror_file = zip_ferror;
  pzlib_filefunc_def->opaque = Stream;
}
```

That was all on the low-level decompression details. Let us get higher into the territory of more friendly abstractions and wrappers. The `clArchiveMountPoint` class wraps an instance of `clArchiveReader` and implements the `CreateReader()`, `FileExists()`, and `MapName()` methods:

```
class clArchiveMountPoint: public iMountPoint
{
public:
  explicit clArchiveMountPoint( const clPtr<ArchiveReader>& R )
  : FReader(R) {}
```

The `CreateReader()` method instantiates a `clMemRawFile` class and attaches an extracted memory block:

```
virtual clPtr<iRawFile> CreateReader(
  const std::string& VirtualName ) const
{
  std::string Name = Arch_FixFileName( VirtualName );
  const void* DataPtr = FReader->GetFileData( Name );
  size_t FileSize = static_cast<size_t>(
    FReader->GetFileSize( Name ) );
  auto File = clMemRawFile::CreateFromManagedBuffer(
    DataPtr, FileSize );
  File->SetFileName( VirtualName );
  File->SetVirtualFileName( VirtualName );
  return File;
}
```

The `FileExists()` method is an indirection to `clArchiveReader::FileExists()`:

```
virtual bool FileExists( const std::string& VirtualName )
  const
{
  return FReader->FileExists(
    Arch_FixFileName( VirtualName ) );
}
```

The `MapName()` implementation is trivial for this type of a mount point:

```
virtual std::string MapName(
  const std::string& VirtualName ) const
{ return VirtualName; }
```

The private section contains only a reference to a `clArchiveReader` object:

```
private:
  clPtr<clArchiveReader> FReader;
};
```

The obvious drawback of the simple `clArchiveMountPoint` is its non-asynchronous blocking implementation. The constructor accepts a fully initialized `clArchiveReader` object, which means we need to block until `clArchiveReader::OpenArchive()` gets its job done. One way to overcome this is to run `OpenArchive()` on a different thread, in a task queue, and create the mount point once the archive is parsed. Of course, all subsequent calls to `CreateReader()` expecting data from this mount point should be postponed until a signal is raised. We encourage the reader to implement this kind of asynchronous mechanism as an exercise using the `clWorkerThread` class discussed in the previous chapter. A more sophisticated archive mount point implementation can accept a constructed `clArchiveReader` and invoke `OpenArchive()` itself. This requires more elaborate architecture as `clFileSystem` and/or `clArchiveMountPoint` classes should have access to a dedicated worker thread. However, it essentially hides all the complexity of time-consuming decompression operations behind the lean interface.

Accessing application assets

To access the data packed inside the `.apk` package on Android in your C++ code, we need to get the path to `.apk` by using Java code and passing the result into our C++ code using JNI.

In the `onCreate()` method, pass the value obtained from `getApplication().getApplicationInfo().sourceDir` into our native code:

```
@Override protected void onCreate( Bundle icicle )
{
  onCreateNative(
    getApplication().getApplicationInfo().sourceDir );
}
public static native void onCreateNative( String APKName );
```

The implementation of `onCreateNative()` can be found in `1_ArchiveFileAccess\jni\Wrappers.cpp` and looks as follows:

```
extern "C"
{
  JNIEXPORT void JNICALL
  Java_com_packtpub_ndkmastering_AppActivity_onCreateNative(
    JNIEnv* env, jobject obj, jstring APKName )
  {
    g_APKName = ConvertJString( env, APKName );
    LOGI( "APKName = %s", g_APKName.c_str() );
    OnStart( g_APKName );
  }
}
```

We use the `ConvertJString()` function to convert `jstring` into `std::string`. The JNI methods `GetStringUTFChars()` and `ReleaseStringUTFChars()` get and release the pointer to an array of UTF8-encoded characters of the string:

```
std::string ConvertJString( JNIEnv* env, jstring str )
{
  if ( !str ) { return std::string(); }
  const jsize len = env->GetStringUTFLength( str );
  const char* strChars = env->GetStringUTFChars(
    str, ( jboolean* )0 );
  std::string Result( strChars, len );
  env->ReleaseStringUTFChars( str, strChars );
  return Result;
}
```

The simple usage example is implemented in the `OnStart()` callback inside the `main.cpp` file. It mounts the path, creates an archive mount point on Android, opens the archive `test.zip` and enlists its content. On a desktop, this code runs and reads `test.zip` which is stored at `assets/test.zip`:

```
void OnStart( const std::string& RootPath )
{
  auto FS = make_intrusive<clFileSystem>();
  FS->Mount( "" );
  FS->Mount( RootPath );
  FS->AddAliasMountPoint( RootPath, "assets" );
  const char* ArchiveName = "test.zip";
  auto File = FS->CreateReader( ArchiveName );
  auto Reader = make_intrusive<clArchiveReader>();
  if ( !Reader->OpenArchive( File ) )
  {
    LOGI( "Bad archive: %s", ArchiveName );
    return;
  }
```

Iterate over all the files in this archive and print their names and contents:

```
for ( size_t i = 0; i != Reader->GetNumFiles(); i++ )
{
  LOGI( "File[%i]: %s", i,
  Reader->GetFileName( i ).c_str() );
  const char* Data =
    reinterpret_cast<const char*>(
      Reader->GetFileDataIdx( i ) );
  LOGI( "Data: %s", std::string( Data,
    static_cast<size_t>(
      Reader->GetFileSizeIdx( i ) ) ).c_str() );
  }
}
```

Check the `1_ArchiveFileAccess` example and try it for yourself. It provides a great debugging experience of the Android file access code on your desktop machine. Use `make all` to build for the desktop environment and `ndk-build & ant debug` to build for Android.

Summary

In this chapter, we learned how to deal with files and `.apk` archives via C++ in a platform-independent way. We will use this functionality in the subsequent chapters to access files.

5
Cross-platform Audio Streaming

In this chapter, we consider the last non-visual component needed to build interactive mobile applications. We are looking for a truly portable implementation of audio playback for Android and desktop PCs. We propose using the OpenAL library, since it is a well-established library on desktop platforms. Audio playback is inherently an asynchronous process, so the decoding and submitting of data to the sound API should be done on a separate thread. We will create an audio streaming library based on the multi-threading code from *Chapter 3, Networking*.

Raw uncompressed audio can take up a lot of memory, so compressed formats of different flavors are used quite often. We consider some of them in this chapter and will show you how to play them in Android using native C++ code and popular third-party libraries.

Initialization and playback

We use the OpenAL cross-platform audio library throughout this chapter. In order to make all the examples simple and self-contained, we start with the minimalistic example which can play a sound from an uncompressed .wav file.

Let's briefly describe what we need to do in order to produce sound. The routines of OpenAL manipulate objects encountered in the playback and recording processes. The ALCdevice object represents a unit of audio hardware. Since multiple threads may produce sound at the same time, another object called ALCcontext is introduced. First, an application opens a device and then a context is created and attached to the opened device. Each context maintains a number of Audio Source objects, because even a single application might need to play multiple sounds simultaneously.

We are getting close to actual sound production. One more object is required as a waveform container, which is called a buffer. An audio recording can be quite lengthy, so we don't submit the entire sound as a single buffer. We read samples in small chunks and submit these chunks using several buffers, usually a couple, into the audio source's queue.

The following pseudo-code depicts how to play a sound which fits entirely into memory:

1. First open a device, create a context, and attach the context to the device.
2. Create an audio source, allocate a single sound buffer.
3. Load waveform data into the buffer.
4. Enqueue the buffer to the audio source.
5. Wait until playback is complete.
6. Destroy the buffer, the source, and the context and close the device.

There is an apparent problem at step 5. We cannot block the UI thread of the application for a couple of seconds, thus the sound playback must be asynchronous. Fortunately, the OpenAL calls are thread-safe and we can perform playback in a separate thread without doing any OpenAL synchronization ourselves.

Let's check the example 1_InitOpenAL. To perform waveform loading in step 3 and keep the code as simple as possible, we take a .wav file and load it into a clBlob object. In step 2, we create an audio source and a buffer with parameters corresponding to those inside the WAV header. Steps 1, 4, and 6 consist only of some OpenAL API calls. Step 5 is done via a busy-wait loop on an atomic conditional variable.

The native C++ entry point for this example starts by creating a separate audio thread declared as a global object g_Sound. The g_FS object contains an instance of the clFileSystem class used to load audio data from files:

```
clSoundThread g_Sound;
clPtr<clFileSystem> g_FS;
int main()
{
  g_FS = make_intrusive<clFileSystem>();
  g_FS->Mount( "." );
  g_Sound.Start();
  g_Sound.Exit( true );
  return 0;
}
```

The `clSoundThread` class contains an OpenAL device and a context. The audio source and buffer handles are also declared for this one-source-one-buffer example:

```
class clSoundThread: public iThread
{
  ALCdevice* FDevice;
  ALCcontext* FContext;
  ALuint FSourceID;
  ALuint FBufferID;
```

The method `Run()` does all the initialization, loading, and finalization:

```
virtual void Run()
{
```

To use OpenAL routines, we should load the library. For Android, Linux, and OS X, the implementation is easy, we just use a statically linked library and that is it. However, for Windows, we load the `OpenAL32.dll` file and retrieve all the necessary function pointers from the dynamic link library:

```
LoadAL();
```

First, we open a device and create a context. The `nullptr` argument for `alcOpenDevice()` means that we are using the default sound device:

```
FDevice = alcOpenDevice( nullptr );
FContext = alcCreateContext( FDevice, nullptr );
alcMakeContextCurrent( FContext );
```

Then we create an audio source and set its volume to the maximum level:

```
alGenSources( 1, &FSourceID );
alSourcef( FSourceID, AL_GAIN, 1.0 );
```

Loading of the waveform, which corresponds to the step 3 in our pseudo-code, is done by reading the entire `.wav` file into the `clBlob` object:

```
auto data = LoadFileAsBlob( g_FS, "test.wav" );
```

The header can be accessed the following way:

```
const sWAVHeader* Header =
  ( const sWAVHeader* )Blob->GetData();
```

We copy bytes from `clBlob` into the sound buffer, skipping the number of bytes corresponding to the size of the header:

```
const unsigned char* WaveData =
  ( const unsigned char* )Blob->GetData() +
  sizeof( sWAVHeader );
PlayBuffer( WaveData, Header->DataSize,
  Header->SampleRate );
```

Now let us just busy-wait for the sound to finish:

```
while ( IsPlaying() ) {}
```

At the end, we stop the source, delete all the objects and unload the OpenAL library:

```
alSourceStop( FSourceID );
alDeleteSources( 1, &FSourceID );
alDeleteBuffers( 1, &FBufferID );
alcDestroyContext( FContext );
alcCloseDevice( FDevice );
UnloadAL();
}
```

The `clSoundThread` class also contains two helper methods. The `IsPlaying()` method checks if the sound is still playing by requesting its state:

```
bool IsPlaying() const
{
  int State;
  alGetSourcei( FSourceID, AL_SOURCE_STATE, &State );
  return State == AL_PLAYING;
}
```

The `PlayBuffer()` method creates a buffer object, fills it with the waveform from the `Data` parameter and starts playback:

```
void PlayBuffer( const unsigned char* Data,
  int DataSize, int SampleRate )
{
  alBufferData( FBufferID, AL_FORMAT_MONO16,
    Data, DataSize, SampleRate );
  alSourcei( FSourceID, AL_BUFFER, FBufferID );
  alSourcei( FSourceID, AL_LOOPING, AL_FALSE );
  alSourcef( FSourceID, AL_GAIN, 1.0f );
  alSourcePlay( FSourceID );
}
```

The preceding code relies on two global functions. `Env_Sleep()` sleeps for a given amount of milliseconds. The Windows version of the code differs slightly from Android and OS X:

```
void Env_Sleep( int Milliseconds )
{
  #if defined(_WIN32)
    Sleep( Milliseconds );
  #elif defined(ANDROID)
    std::this_thread::sleep_for(
      std::chrono::milliseconds( Milliseconds ) );
  #else
    usleep( static_cast<useconds_t>( Milliseconds ) * 1000 );
  #endif
}
```

 We use `Sleep()` with Windows in order to be compatible with some MinGW distros that lack support for `std::chrono`. If you want to use Visual Studio, just stick to `std::this_thread::sleep_for()`.

The `LoadFileAsBlob()` function uses the provided `clFileSystem` object to load the contents of a file into the memory block. We reuse this routine in most of our subsequent code samples.

```
clPtr<clBlob> LoadFileAsBlob(
  const clPtr<clFileSystem>& FileSystem,
  const std::string& Name )
{
  auto Input = FileSystem->CreateReader( Name );
  auto Res = make_intrusive<clBlob>();
  Res->AppendBytes( Input->MapStream(), Input->GetSize() );
  return Res;
}
```

If you compile and run this example on a desktop machine by typing `make all`, you should hear a short ding sound. Let us proceed further and learn how to do sound streaming before we end up with an Android application.

Streaming sounds

Now that we can play short audio samples, it is time to organize our audio system into classes and take a closer look at the 2_Streaming example. Long audio samples, such as background music, require a lot of memory in a decompressed form. Streaming is a technique to decompress them in small chunks, piece by piece. The clAudioThread class manages the initialization and does everything from the previous sample except playing the sound:

```
class clAudioThread: public iThread
{
public:
  clAudioThread()
  : FDevice( nullptr )
  , FContext( nullptr )
  , FInitialized( false )
  {}
  virtual void Run()
  {
    if ( !LoadAL() ) { return; }
    FDevice = alcOpenDevice( nullptr );
    FContext = alcCreateContext( FDevice, nullptr );
    alcMakeContextCurrent( FContext );
    FInitialized = true;
    while ( !IsPendingExit() ) { Env_Sleep( 100 ); }
    alcDestroyContext( FContext );
    alcCloseDevice( FDevice );
    UnloadAL();
  }
```

This method is used to synchronize the start of the audio thread with its users:

```
  virtual void WaitForInitialization() const
  {
    while ( !FInitialized ) {}
  }
private:
  std::atomic<bool> FInitialized;
  ALCdevice* FDevice;
  ALCcontext* FContext;
};
```

The clAudioSource class represents a single sound producing entity. The wave data is not stored in the source itself and we postpone the description of the clAudioSource class. Right now, we introduce the iWaveDataProvider interface class which provides the data for the next audio buffer. The reference to an iWaveDataProvider instance is stored in the clAudioSource class:

```
class iWaveDataProvider: public iIntrusiveCounter
{
public:
```

The audio signal properties are stored in these three fields:

```
int FChannels;
int FSamplesPerSec;
int FBitsPerSample;
iWaveDataProvider()
: FChannels( 0 )
, FSamplesPerSec( 0 )
, FBitsPerSample( 0 ) {}
```

Two pure virtual methods give access to the current wave data played by the audio source. They are to be implemented in the actual decoder subclasses:

```
virtual unsigned char* GetWaveData() = 0;
virtual size_t GetWaveDataSize() const = 0;
```

The IsStreaming() method tells us if this provider represents a continuous stream or a single chunk of audio data, like the one in the previous example. The StreamWaveData() method loads, decodes, or generates values in the buffer accessed by the GetWaveData() function; it is usually implemented in subclasses as well. When clAudioSource needs more audio data to enqueue into a buffer, it invokes the StreamWaveData() method:

```
virtual bool IsStreaming() const { return false; }
virtual int StreamWaveData( int Size ) { return 0; }
```

The last auxiliary function returns the internal data format used by OpenAL. Here we support only stereo and mono signals with 8 or 16 bits per sample:

```
ALuint GetALFormat() const
{
  if ( FBitsPerSample == 8 )
    return ( FChannels == 2 ) ?
      AL_FORMAT_STEREO8 : AL_FORMAT_MONO8;
  if ( FBitsPerSample == 16 )
```

```
      return ( FChannels == 2 ) ?
        AL_FORMAT_STEREO16 : AL_FORMAT_MONO16;
    return AL_FORMAT_MONO8;
  }
};
```

Our basic sound decoding is done in the `clStreamingWaveDataProvider` class. It contains the `FBuffer` data vector and the number of useful bytes in it:

```
class clStreamingWaveDataProvider: public iWaveDataProvider
{
public:
  clStreamingWaveDataProvider()
  : FBufferUsed( 0 )
  {}
  virtual bool IsStreaming() const override
  { return true; }
  virtual unsigned char* GetWaveData() override
  { return ( unsigned char* )&FBuffer[0]; }
  virtual size_t GetWaveDataSize() const override
  { return FBufferUsed; }
  std::vector<char> FBuffer;
  size_t FBufferUsed;
};
```

We are ready to describe the class `clAudioSource` which does the actual heavy lifting. The constructor creates an OpenAL audio source object, sets the volume level and disables looping:

```
class clAudioSource: public iIntrusiveCounter
{
public:
  clAudioSource()
  : FWaveDataProvider( nullptr )
  , FBuffersCount( 0 )
  {
    alGenSources( 1, &FSourceID );
    alSourcef( FSourceID, AL_GAIN, 1.0 );
    alSourcei( FSourceID, AL_LOOPING, AL_FALSE );
  }
```

We have two different use cases. If the attached `iWaveDataProvider` supports streaming, we need to create and maintain at least two sound buffers. Both buffers are enqueued to the OpenAL playback queue and swapped as soon as one of them finishes playing. At each swapping event, we call the `StreamWaveData()` method of `iWaveDataProvider` to stream the data into the next audio buffer. If the `iWaveDataProvider` is not streamed, we only need a single buffer which is initialized at the beginning.

The `Play()` method fills both buffers with decoded data and calls `alSourcePlay()` to start playback:

```
void Play()
{
  if ( IsPlaying() ) { return; }
  if ( !FWaveDataProvider ) { return; }
  int State;
  alGetSourcei( FSourceID, AL_SOURCE_STATE, &State );
  if ( State != AL_PAUSED &&
    FWaveDataProvider->IsStreaming() )
  {
    UnqueueAll();
    StreamBuffer( FBufferID[0], BUFFER_SIZE );
    StreamBuffer( FBufferID[1], BUFFER_SIZE );
    alSourceQueueBuffers( FSourceID, 2, &FBufferID[0] );
  }
  alSourcePlay( FSourceID );
}
```

The `Stop()` and `Pause()` methods call appropriate OpenAL routines to stop and pause playback respectively:

```
void Stop()
{
  alSourceStop( FSourceID );
}
void Pause()
{
  alSourcePause( FSourceID );
  UnqueueAll();
}
```

The `LoopSound()` and `SetVolume()` methods control playback parameters:

```
void LoopSound( bool Loop )
{
alSourcei( FSourceID, AL_LOOPING, Loop ? 1 : 0 );
}
void SetVolume( float Volume )
{
  alSourcef( FSourceID, AL_GAIN, Volume );
}
```

The `IsPlaying()` method is copied from the previous example:

```
bool IsPlaying() const
{
  int State;
  alGetSourcei( FSourceID, AL_SOURCE_STATE, &State );
  return State == AL_PLAYING;
}
```

The `StreamBuffer()` method writes the newly generated audio data into one of the buffers:

```
int StreamBuffer( unsigned int BufferID, int Size )
{
  int ActualSize =
    FWaveDataProvider->StreamWaveData( Size );
  alBufferData( BufferID,
    FWaveDataProvider->GetALFormat(),
    FWaveDataProvider->GetWaveData(),
    ( int )FWaveDataProvider->GetWaveDataSize(),
    FWaveDataProvider->FSamplesPerSec );
  return ActualSize;
}
```

The `Update()` method should be called often enough to prevent audio buffers from underflow. However, this method only matters if the attached `iWaveDataProvider` represents an audio stream:

```
void Update( float DeltaSeconds )
{
  if ( !FWaveDataProvider ) { return; }
  if ( !IsPlaying() ) { return; }
  if ( FWaveDataProvider->IsStreaming() )
  {
```

We ask OpenAL how many buffers have been processed:

```
int Processed;
alGetSourcei( FSourceID, AL_BUFFERS_PROCESSED,
  &Processed );
```

We remove each processed buffer from the queue and call `StreamBuffer()` to decode more data. Finally, we readd the buffer to the playback queue:

```
while ( Processed-- )
{
  unsigned int BufID;
  alSourceUnqueueBuffers(
    FSourceID, 1, &BufID );
  StreamBuffer( BufID, BUFFER_SIZE );
  alSourceQueueBuffers(
    FSourceID, 1, &BufID );
}
}
}
```

The destructor stops playback and destroys the OpenAL audio source and buffers:

```
virtual ~clAudioSource()
{
  Stop();
  alDeleteSources( 1, &FSourceID );
  alDeleteBuffers( FBuffersCount, &FBufferID[0] );
}
```

The `BindWaveform()` method attaches a new `iWaveDataProvider` to this audio source instance:

```
void BindWaveform( clPtr<iWaveDataProvider> Wave )
{
  FWaveDataProvider = Wave;
  if ( !Wave ) { return; }
```

For a streaming `iWaveDataProvider`, we need two buffers. One is being played while the other is being updated:

```
if ( FWaveDataProvider->IsStreaming() )
{
  FBuffersCount = 2;
  alGenBuffers( FBuffersCount, &FBufferID[0] );
}
else
```

If the attached waveform is not a stream or, more specifically, it is not compressed, we create a single buffer and copy all the data into it:

```
    {
      FBuffersCount = 1;
      alGenBuffers( FBuffersCount, &FBufferID[0] );
      alBufferData( FBufferID[0],
        FWaveDataProvider->GetALFormat(),
        FWaveDataProvider->GetWaveData(),
        ( int )FWaveDataProvider->GetWaveDataSize(),
        FWaveDataProvider->FSamplesPerSec );
      alSourcei( FSourceID, AL_BUFFER, FBufferID[0] );
    }
  }
```

The private `UnqueueAll()` method uses `alSourceUnqueueBuffers()` to clear the OpenAL playback queue:

```
private:
  void UnqueueAll()
  {
    int Queued;
    alGetSourcei( FSourceID, AL_BUFFERS_QUEUED, &Queued );
    if ( Queued > 0 )
    {
      alSourceUnqueueBuffers( FSourceID, Queued, &FBufferID[0] );
    }
  }
```

The tail part of this class defines the reference to the attached `iWaveDataProvider`, internal handles of OpenAL objects, and the number of allocated buffers:

```
  clPtr<iWaveDataProvider> FWaveDataProvider;
  unsigned int FSourceID;
  unsigned int FBufferID[2];
  int FBuffersCount;
};
```

To demonstrate some basic streaming capabilities, we change the sample code from `1_InitOpenAL` and create an audio source with the attached tone generator, described in the following code:

```
  class clSoundThread: public iThread
  {
```

```
    virtual void Run()
    {
      g_Audio.WaitForInitialization();
      auto Src = make_intrusive<clAudioSource>();
      Src->BindWaveform( make_intrusive<clToneGenerator>() );
      Src->Play();
      double Seconds = Env_GetSeconds();
      while ( !IsPendingExit() )
      {
        float DeltaSeconds =
          static_cast<float>( Env_GetSeconds() - Seconds );
        Src->Update( DeltaSeconds );
        Seconds = Env_GetSeconds();
      }
    }
  };
```

In this example, we deliberately avoid the problem of decompressing the sound to focus on the streaming logic. So we start with a procedurally generated sound. The `clToneGenerator` class overrides the `StreamWaveData()` method and generates a sine wave, or a pure tone. To avoid audible glitches, we have to sample the sine function carefully and remember the integer index of the last generated sample. This index is stored in the `FLastOffset` field and used in calculations in each iteration.

The constructor of the class sets audio parameters to 16-bit 44.1kHz and allocates some space in the `FBuffer` container. The base frequency of this tone is set to 440 Hz:

```
class clToneGenerator : public clStreamingWaveDataProvider
{
public:
  clToneGenerator()
  : FFrequency( 440.0f )
  , FAmplitude( 350.0f )
  , FLastOffset( 0 )
  {
    FBufferUsed = 100000;
    FBuffer.resize( 100000 );
    FChannels = 2;
    FSamplesPerSec = 44100;
    FBitsPerSample = 16;
  }
```

In `StreamWaveData()`, we check for available space in the `FBuffer` vector and reallocate it if necessary:

```
virtual int StreamWaveData( int Size )
{
  if ( Size > static_cast<int>( FBuffer.size() ) )
  {
    FBuffer.resize( Size );
    LastOffset = 0;
  }
```

Finally, we calculate the audio sample. The frequency is recalculated based on the sample count:

```
const float TwoPI = 2.0f * 3.141592654f;
float Freq = TwoPI * FFrequency /
  static_cast<float>( FSamplesPerSec );
```

Since we need `Size` bytes and our signal contains two channels with 16-bit samples, we need a total of `Size/4` samples:

```
for ( int i = 0 ; i < Size / 4 ; i++ )
{
  float t = Freq * static_cast<float>(
    i + LastOffset );
  float val = FAmplitude * std::sin( t );
```

We convert the floating point value into the 16-bit signed integer and put low and high bytes of this integer into `FBuffer`. For each channel, we store two bytes:

```
short V = static_cast<short>( val );
FBuffer[i * 4 + 0] = V & 0xFF;
FBuffer[i * 4 + 1] = V >> 8;
FBuffer[i * 4 + 2] = V & 0xFF;
FBuffer[i * 4 + 3] = V >> 8;
}
```

After the calculation, we increment the sample count and take the remainder to avoid integer overflow in the counter:

```
  LastOffset += Size / 4;
  LastOffset %= FSamplesPerSec;
  return ( FBufferUsed = Size );
}
float FFrequency;
```

```
    float FAmplitude;
private:
    int LastOffset;
};
```

The compiled example will produce a pure tone of 440 Hz. We encourage you to
change the value of `clToneGenerator::FFrequency` and see how it works. You
can even create a simple tuning fork application for your musical instruments using
this example. As for musical instruments, let's generate some audio data to mimic a
stringed musical instrument.

Stringed musical instrument model

Let's implement a simple physical model of a stringed musical instrument using the
code from the previous example. Later you can use these routines to create a small
interactive synthesizer for Android.

The string is modelled as a sequence of point masses oscillating vertically. Strictly
speaking, we solve the linear one-dimensional wave equation with certain initial and
boundary conditions. The sound is produced by taking values of the solution at the
position of sound pickup.

We need the `clGString` class to store all the model values and the final result. The
method `GenerateSound()` precalculates string parameters and resizes the data
containers accordingly:

```
class clGString
{
public:
    void GenerateSound()
    {
        // 4 seconds, 1 channel, 16 bit
        FSoundLen  = 44100 * 4 * 2;
        FStringLen = 200;
```

The `Frc` value is the normalized fundamental frequency of the sound. Overtones are
implicitly created by the physical model:

```
        float Frc = 0.5f;
        InitString( Frc );
        FSamples.resize( FsoundLen );
        FSound.resize( FsoundLen );
        float MaxS = 0;
```

After the initialization stage, we perform integration of the wave equation by calling the `Step()` method in a loop. The `Step()` member function returns the displacement of the string at the pickup position:

```
for ( int i = 0; i < FSoundLen; i++ )
{
   FSamples[i] = Step();
```

At each step, we clamp the value to the maximum:

```
   if ( MaxS < fabs(FSamples[i]) )
   MaxS = fabs( FSamples[i] );
}
```

Finally, we convert floating point values to signed short integers. To avoid overflows, each sample is divided by the `MaxS` value:

```
   const float SignedShortMax = 32767.0f;
   float k = SignedShortMax / MaxS;
   for ( int i = 0; i < FSoundLen; i++ )
   {
      FSound [i] = FSamples [i] * k;
   }
}
std::vector<short int> FSound;
private:
int FPickPos;
int FSoundLen;
std::vector<float> FSamples;
std::vector<float> FForce;
std::vector<float> FVel;
std::vector<float> FPos;
float k1, k2;
int FStringLen;
void InitString(float Freq)
{
   FPos.resize(FStringLen);
   FVel.resize(FStringLen);
   FForce.resize(FStringLen);
   const float Damping = 1.0f / 512.0f;
   k1 = 1 - Damping;
   k2 = Damping / 2.0f;
```

We place the sound pickup closer to the end:

```
FPickPos = FStringLen * 5 / 100;
for ( int i = 0 ; i < FStringLen ; i++ )
{
   FVel[i] = FPos[i] = 0;
}
```

For better results, we produce a slight variation in the mass of a string element:

```
for ( int i = 1 ; i < FStringLen - 1 ; i++ )
{
   float m = 1.0f + 0.5f * (frand() - 0.5f);
   FForce[i] = Freq / m;
}
```

At the beginning, we set non-zero velocities for the second part of the string:

```
for ( int i = FStringLen/2; i < FStringLen - 1; i++ )
{
   FVel[i] = 1;
}
}
```

The frand() member function returns a pseudo-random floating point value in the 0..1 range:

```
inline float frand()
{
   return static_cast<float>( rand() ) /
      static_cast<float>( RAND_MAX );
}
```

 The usage of std::random is the preferred way of getting a pseudo-random number if your compiler supports it.

Here is the the way to generate a pseudo-random floating point number uniformly distributed in the range 0...1 using the new C++11 Standard Library:

```
std::random_device rd;
std::mt19937 gen( rd() );
std::uniform_real_distribution<> dis( 0.0, 1.0 );
float frand()
{
   return static_cast<float>( dis( gen ) );
}
```

Though this short code snippet is not used in our source code bundle, it may be of use to you. Let us return to the code of our example.

The `Step()` method makes a single step and integrates the equations of the string motion. At the end of the step, a value from the `FPos` vector at the `FPickPos` position is taken as the next sample of the sound. For readers familiar with numerical methods, it might seem strange that there is no time step specification, implicitly it is 1/44100th of a second:

```
float Step()
{
```

At first, we enforce boundary conditions, those are fixed endpoints at both ends of the string:

```
FPos[0] = FPos[FStringLen - 1] = 0;
FVel[0] = FVel[FStringLen - 1] = 0;
```

According to Hooke's law (http://en.wikipedia.org/wiki/Hooke's_law), the force is proportional to the extension:

```
for ( int i = 1 ; i < FStringLen - 1 ; i++ )
{
    float d = (FPos[i - 1] + FPos[i + 1]) * 0.5f - FPos[i];
    FVel[i] += d * FForce[i];
}
```

To ensure numerical stability, we apply some artificial damping and take the average of the neighboring velocities. Failing to do so produces some unwanted tinkling sound:

```
for ( int i = 1 ; i < FStringLen - 1 ; i++ )
{
    FVel[i] = FVel[i] * k1 +
        (FVel[i - 1] + FVel[i + 1]) * k2;
}
```

Finally, we update positions:

```
for ( int i = 1 ; i < FStringLen ; i++ )
{
    FPos[i] += FVel[i];
}
```

To record our sound, we take only one position of the string:

```
    return FPos[FPickPos];
  }
};
```

The 1_InitOpenAL example is easily modified to generate a string sound instead of loading a .wav file. We create the clGString instance and call the GenerateSound() method. After that, we get the FSound vector and submit it to the PlayBuffer() method of an audio source:

```
clGString String;
String.GenerateSound();
const unsigned char* Data =
  (const unsigned char*)&String.FSound[0];
PlayBuffer( Data, (int)String.FSound.size() );
```

Here, the sampling rate is hardcoded at 44100 Hz. Try the 3_GuitarStringSound example for the complete code and hear it for yourself. Note that the startup time may be a bit high due to heavy pre-calculations before the sound can be played. However, the code is very simple and we leave it as an exercise to the reader to compile it for Android, taking all the necessary makefiles and wrappers from the subsequent examples. And in the meantime, we will do the stuff that can be run on Android out-of-the-box.

Decoding compressed audio

Now that we have implemented the basic audio streaming system, it is time to use a couple of third-party libraries to read compressed audio files. Basically, what we need to do is to override the StreamWaveData() function inside the clStreamingWaveDataProvider class. This function, in turn, calls the ReadFromFile() method where the actual decoding is done. The initialization of the decoder is done in the constructor and, for the abstract iDecodingProvider class, we only store the reference to a data blob. All the compressed data for the file is stored in a clBlob object:

```
class iDecodingProvider: public StreamingWaveDataProvider
{
protected:
  virtual int ReadFromFile( int Size, int BytesRead ) = 0;
  clPtr<clBlob> FRawData;
public:
  bool FLoop;
```

```
bool FEof;
iDecodingProvider( const clPtr<clBlob>& Blob )
: FRawData( Blob )
, FLoop( false )
, FEof( false )
{}
virtual bool IsEOF() const { return FEof; }
```

The `StreamWaveData()` method does the job of decoding. The first few lines ensure there is enough space in `FBuffer` to contain the decoded data:

```
virtual int StreamWaveData( int Size ) override
{
  int OldSize = ( int )FBuffer.size();
  if ( Size > OldSize )
  {
```

After the buffer is reallocated, we fill new bytes with zeroes, because non-zero values can produce unexpected noise:

```
    FBuffer.resize( Size, 0 );
  }
  if ( FEof ) { return 0; }
```

Since `ReadFromFile()` may return insufficient data, we call it in a loop incrementing the number of bytes read:

```
  int BytesRead = 0;
  while ( BytesRead < Size )
  {
    int Ret = ReadFromFile( Size, BytesRead );
    if ( Ret > 0 ) BytesRead += Ret;
```

The return value of zero from `ReadFromFile()` means we have reached the end of the stream:

```
    else if ( Ret == 0 )
    {
      FEof = true;
```

Looping is done by calling `Seek()` and setting the `FEof` flag:

```
if ( FLoop )
{
  Seek( 0 );
  FEof = false;
  continue;
}
break;
}
```

A negative value in `Ret` indicates a reading error has occurred. We stop decoding in this case:

```
else
{
  Seek( 0 );
  FEof = true;
  break;
}
}
return ( FBufferUsed = BytesRead );
}
};
```

The next two sections show how to decode different formats of audio files using popular third-party libraries.

Decoding tracker music using the ModPlug library

The first library we will tackle to decode audio files is the ModPlug library by Olivier Lapicque. Most popular tracker music file formats `http://en.wikipedia.org/wiki/Module_file` can be decoded and converted into the waveform suitable for OpenAL using ModPlug. We will introduce the `clModPlugProvider` class which implements the `ReadFromFile()` routine. The constructor of the class loads the memory blob into the `ModPlugFile` object and assigns default audio parameters:

```
class clModPlugProvider: public iDecodingProvider
{
private:
  ModPlugFile* FModFile;
```

```
public:
  ModPlugProvider( const clPtr<clBlob>& Blob ):
  {
    DecodingProvider( Blob )
    FChannels = 2;
    FSamplesPerSec = 44100;
    FBitsPerSample = 16;
    FModFile = ModPlug_Load_P(
      ( const void* )
      FRawData->GetDataConst(), ( int )FRawData->GetSize()
    );
  }
```

The destructor cleans up ModPlug:

```
virtual ~ModPlugProvider() { ModPlug_Unload_P( FModFile ); }
```

The `ReadFromFile()` method calls `ModPlug_Read()` to fill `FBuffer`:

```
virtual int ReadFromFile( int Size, int BytesRead )
{
  return ModPlug_Read_P( FModFile,
    &FBuffer[0] + BytesRead, Size - BytesRead );
}
```

Stream positioning is done using the `ModPlug_Seek()` routine. Inside the ModPlug API, all the timing is done in milliseconds:

```
virtual void Seek( float Time )
{
  FEof = false;
  ModPlug_Seek_P( FModFile, ( int )( Time * 1000.0f ) );
}
};
```

To use this wave data provider, we attach its instance to a `clAudioSource` object:

```
Src->BindWaveform(
  make_intrusive<clModPlugProvider>(
    LoadFileAsBlob( g_FS, "augmented_emotions.xm" )
  )
);
```

Other details are reused from our previous examples. The `4_ModPlug` folder can be built and run on Android as well as on Windows. Use `ndk-build` and `ant debug` to create `.apk` for Android, and `make all` to create a Windows executable.

Decoding MP3 files

Most of the patents for the MPEG-1 Layer 3 format expire by the end of the year 2015, so it is worth mentioning the MiniMP3 library by Fabrice Bellard. Using this library is not harder than ModPlug, because we have already done all the grunt work in `iDecodingProvider`. Let's take a look at the 5_MiniMP3 example. The `clMP3Provider` class creates the decoder instance and reads stream parameters by reading some frames from the beginning:

```
class clMP3Provider: public iDecodingProvider
{
public:
  clMP3Provider( const clPtr<clBlob>& Blob )
  : iDecodingProvider( Blob )
  {
    FBuffer.resize(MP3_MAX_SAMPLES_PER_FRAME * 8);
    FBufferUsed = 0;
    FBitsPerSample = 16;
    mp3 = mp3_create();
    bytes_left = ( int )FRawData->GetSize();
```

At the beginning, we set the stream position to the beginning of the `clBlob` object:

```
    stream_pos = 0;
    byte_count = mp3_decode((mp3_decoder_t*)mp3,
      ( void* )FRawData->GetData(), bytes_left,
      (signed short*)&FBuffer[0], &info);
    bytes_left -= byte_count;
```

We need the information about the audio data, so we fetch it from the `info` structure:

```
    FSamplesPerSec = info.sample_rate;
    FChannels = info.channels;
  }
```

There is nothing special in the destructor, here it is:

```
  virtual ~MP3Provider()
  {
    mp3_done( &mp3 );
  }
```

The `ReadFromFile()` method keeps track of the bytes left in the source stream and fills the `FBuffer` container. Both the constructor and this method use the `bytes_left` and `stream_pos` fields to keep the current stream position and the remaining number of bytes:

```
virtual int ReadFromFile( int Size, int BytesRead )
{
  byte_count = mp3_decode( (mp3_decoder_t*)mp3,
    (( char* )FRawData->GetData()) + stream_pos, bytes_left,
    (signed short *)(&FBuffer[0] + BytesRead), &info);
  bytes_left -= byte_count;
  stream_pos += byte_count;
  return info.audio_bytes;
}
```

Seeking is not so obvious with variable bitrate streams, so we leave this implementation as an exercise for the interested reader. The simplest case with a fixed bitrate should just recalculate `Time` from seconds to sample rate units and then set the `stream_pos` variable:

```
virtual void Seek( float Time ) override
{
  FEof = false;
}
private:
  mp3_decoder_t mp3;
  mp3_info_t info;
  int stream_pos;
  int bytes_left;
  int byte_count;
};
```

To use it, we attach the provider to a `clAudioSource` object, just like with ModPlug:

```
Src->BindWaveform( make_intrusive<clMP3Provider>(
  LoadFileAsBlob( g_FS, "test.mp3" ) ) );
```

Again, this example is runnable on Android, just go and try it.

 This code does not deal properly with some ID3 tags. If you want to write a generic music player based on our code, refer to this open source project written by the authors: `https://github.com/corporateshark/PortAMP`.

Decoding OGG files

There is yet another popular audio format worth mentioning. Ogg Vorbis is a completely open, patent-free, professional audio encoding and streaming technology with all the benefits of Open Source `http://www.vorbis.com`. The big picture of the OGG decoding and playback process is similar to that of MP3. Let's take a look at the example 6_OGG. The `Decoders.cpp` file is extended with definitions of the OGG Vorbis functions `OGG_clear_func()`, `OGG_open_callbacks_func()`, `OGG_time_seek_func()`, `OGG_read_func()`, `OGG_info_func()`, and `OGG_comment_func()`. The functions are linked against a static library on Android or loaded from a `.dll` file on Windows. The main difference compared to MiniMP3 API is in providing a set of data reading callbacks to the OGG decoder. These callbacks are implemented in the `OGG_Callbacks.inc` file. The `OGG_ReadFunc()` callback reads data into the decoder:

```
static size_t OGG_ReadFunc(
  void* Ptr, size_t Size, size_t NMemB, void* DataSource )
{
  clOggProvider* OGG =
    static_cast<clOggProvider*>( DataSource );
  size_t DataSize = OGG->FRawData->GetSize();
  ogg_int64_t BytesRead = DataSize - OGG->FOGGRawPosition;
  ogg_int64_t BytesSize = Size * NMemB;
  if ( BytesSize < BytesRead ) { BytesRead = BytesSize; }
```

It is based on our filesystem abstraction and memory mapped files:

```
memcpy(Ptr,
  ( unsigned char* )OGG->FRawData->GetDataConst() +
  OGG->FOGGRawPosition, ( size_t )BytesRead );
OGG->FOGGRawPosition += BytesRead;
return ( size_t )BytesRead;
}
```

The `OGG_SeekFunc()` callback seeks the input stream using different relative positioning modes:

```
static int OGG_SeekFunc( void* DataSource, ogg_int64_t Offset,
  int Whence )
{
  clOggProvider* OGG =
    static_cast<clOggProvider*>( DataSource );
  size_t DataSize = OGG->FRawData->GetSize();
  if ( Whence == SEEK_SET )
```

```
    {
      OGG->FOGGRawPosition = Offset;
    }
    else if ( Whence == SEEK_CUR )
    {
      OGG->FOGGRawPosition += Offset;
    }
    else if ( Whence == SEEK_END )
    {
      OGG->FOGGRawPosition = DataSize + Offset;
    }
    if ( OGG->FOGGRawPosition > ( ogg_int64_t )DataSize )
    {
      OGG->FOGGRawPosition = ( ogg_int64_t )DataSize;
    }
    return static_cast<int>( OGG->FOGGRawPosition );
  }
```

The `OGG_CloseFunc()` and `OGG_TellFunc()` functions are trivial:

```
    static int OGG_CloseFunc( void* DataSource )
    {
      return 0;
    }
     static long OGG_TellFunc( void* DataSource )
    {
     return static_cast<int>(
        (( clOggProvider* )DataSource )->FOGGRawPosition );
    }
```

These callbacks are used in the constructor of `clOggProvider` to set up the decoder:

```
    clOggProvider( const clPtr<clBlob>& Blob )
    : iDecodingProvider( Blob )
    , FOGGRawPosition( 0 )
    {
      ov_callbacks Callbacks;
      Callbacks.read_func  = OGG_ReadFunc;
      Callbacks.seek_func  = OGG_SeekFunc;
      Callbacks.close_func = OGG_CloseFunc;
      Callbacks.tell_func  = OGG_TellFunc;
      OGG_ov_open_callbacks(
        this, &FVorbisFile, nullptr, -1, Callbacks );
```

The stream parameters, like the number of channels, sampling rate and bits per sample, are retrieved here:

```
vorbis_info* VorbisInfo =
   OGG_ov_info ( &FVorbisFile, -1 );
FChannels = VorbisInfo->channels;
FSamplesPerSec = VorbisInfo->rate;
FBitsPerSample = 16;
}
```

The destructor is trivial:

```
virtual ~clOggProvider()
{
   OGG_ov_clear( &FVorbisFile );
}
```

The ReadFromFile() and Seek() methods are pretty much similar in spirit to what we have done when dealing with MiniMP3:

```
virtual int ReadFromFile( int Size, int BytesRead ) override
{
   return ( int )OGG_ov_read(
      &FVorbisFile, &FBuffer[0] + BytesRead, Size - BytesRead,
      0, FBitsPerSample / 8, 1,
      &FOGGCurrentSection );
}
virtual void Seek( float Time ) override
{
   FEof = false;
   OGG_ov_time_seek( &FVorbisFile, Time );
}
private:
```

This is the place where the callbacks mentioned in the preceding section are defined. Of course, they can be defined in-place instead of moving them into a separate file. However, we find this kind of separation more logical for this example; separating logically, the data provider concept and OGG Vorbis related APIs:

```
#include "OGG_Callbacks.inc"
OggVorbis_File FVorbisFile;
ogg_int64_t FOGGRawPosition;
int FOGGCurrentSection;
};
```

This example is also Android-capable out-of-the-box. Run the following command to get the `.apk` on your device:

```
>ndk-build
>ant debug
>adb install -r bin/App1-debug.apk
```

Now start the activity and enjoy the music! In subsequent chapters we will add more interesting audio stuff on top of the material from this chapter.

Summary

In this chapter, we learned how to play audio on Android using portable C++ code and open source third-party libraries. The provided examples are capable of playing `.mp3`, and `.ogg` audio files along with `.it`, `.xm`, `.mod`, and `.s3m` modules. We also learnt how to generate your own waveforms to simulate musical instruments. The code is portable across many systems and can be run and debugged on Android and Windows. Now, once we are done with audio, it is time to proceed to the next chapter and render some graphics with OpenGL.

6

OpenGL ES 3.1 and Cross-platform Rendering

In this chapter, we will learn how to implement an abstraction layer on top of OpenGL 4 and OpenGL ES 3 in order to make our graphics applications runnable on Android and desktop machines. Let's start with some basic vector and linear algebra classes.

Linear algebra and transformations

In the `Core/VecMath.h` file, there is a bunch of vector and matrix specific classes and helpers. The main classes we use are `LVector2`, `LVector3`, `LVector4`, `LMatrix3`, `LMatrix4`, and `LQuaternion` for which basic algebraic operations are defined. There are shortcuts for them to make writing of any math-heavy code easier:

```
using vec2 = LVector2;
using vec3 = LVector3;
using vec4 = LVector4;
using mat3 = LMatrix3;
using mat4 = LMatrix4;
using quat = LQuaternion;
```

This tiny math library is basically a tight squeeze of some algebra code from Linderdaum Engine (http://www.linderdaum.com).

Besides this, there is a set of useful functions in the namespace `Math` dealing with different projection transformations calculation. They will be heavily used in the subsequent chapters.

Graphics initialization using SDL2

In our previous book, *Android NDK Game Development Cookbook, Packt Publishing*, we learned in great detail how to initialize OpenGL ES 2 on Android and OpenGL 3 Core Profile on desktop using our own handcrafted code. Now, we will do it using the SDL2 library, which is available at `https://www.libsdl.org`. Let's take a look at the 1_GLES3 example. The Java code for this example, besides SDL2 internals of course, is short and simple:

```
package com.packtpub.ndkmastering;
import android.app.Activity;
import android.os.Bundle;
public class AppActivity extends org.libsdl.app.SDLActivity
{
  static
  {
    System.loadLibrary( "NativeLib" );
  }
  public static AppActivity m_Activity;
  @Override protected void onCreate( Bundle icicle )
  {
    super.onCreate( icicle );
    m_Activity = this;
  }
};
```

Everything else is done in the C++ code. There is the `main()` function, which is redefined by SDL2 using a macro to make our application look like a desktop one:

```
int main(int argc, char* argv[])
{
  clSDL SDLLibrary;
```

First, a window and an OpenGL rendering context are created using the `clSDLWindow` class:

```
  g_Window = clSDLWindow::CreateSDLWindow( "GLES3", 1024, 768 );
```

Then, we can retrieve pointers to OpenGL functions. This abstraction is superior to statically linking against an OpenGL library since it makes our code more portable. For example, you cannot statically link to core OpenGL functions on Windows without using third-party libraries:

```
  LGL3 = std::unique_ptr<sLGLAPI>( new sLGLAPI() );
  LGL::GetAPI( LGL3.get() );
```

This is the callback we already used in *Chapter 4, Organizing a Virtual Filesystem*, when dealing with virtual filesystems. We won't need any path in this example, so let's just use an empty string:

```
OnStart( "" );
```

The event loop is done explicitly and contains a call to the OnDrawFrame() function:

```
while( g_Window && g_Window->HandleInput() )
{
  OnDrawFrame();
  g_Window->Swap();
}
g_Window = nullptr;
return 0;
}
```

These wrapper classes (clSDL and clSDLWindow) are declared in files SDLLibrary.h and SDLWindow.h respectively. The clSDL class is a RAII wrapper on top of SDL and does initialization and deinitialization of the library in its constructor and destructor:

```
clSDL()
{
  SDL_Init( SDL_INIT_VIDEO );
}
virtual ~clSDL()
{
  SDL_Quit();
}
```

The clSDLWindow class represents a window abstraction with an OpenGL context and a system message pump attached to it:

```
class clSDLWindow: public iIntrusiveCounr
{
private:
  SDL_Window* m_Window;
  SDL_GLContext m_Context;
  float m_Width;
  float m_Height;
  std::atomic<bool> m_Pendingit;
public:
  clSDLWindow(
    const std::string& Title, int Width, int Height );
```

```
virtual ~clSDLWindow();
void RequestExit()
{
  m_PendingExit = true;
}
void Swap();
```

This member function performs one iteration of the message loop:

```
bool HandleInput()
{
  SDL_Event Event;
  while ( SDL_PollEvent(&Event) && !m_PendingExit )
  {
    if ( (Event.type == SDL_QUIT) ||
      !this->HandleEvent( Event ) )
    m_PendingExit = true;
  }
  return !m_PendingExit;
}
```

Convert integer coordinates into floating point normalized coordinates 0..1 to make it easier to use screens of different resolutions:

```
vec2 GetNormalizedPoint( int x, int y ) const
{
  return vec2(
    static_cast<float>(x) / m_Width,
    static_cast<float>(y) / m_Height
  );
}
```

The following method is useful to construct a projection matrix for the current window:

```
float GetAspect() const
{
  return m_Width / m_Height;
}
```

A public static helper method to create an instance of clSDLWindow is as foows:

```
public:
  static clPtr<clSDLWindow> CreateSDLWindow(
    const std::string& Title, int Width, int Height )
```

```
  {
    return make_intrusive<clSDLWindow>(
      Title, Width, Height );
  }
```

The `HandleEvent()` member function does the job of dispatching an SDL2 event to our callcks:

```
private:
  bool HandleEvent( const SDL_Event& Event );};
```

The implementation of `HandleEvent()` is as follows:

```
bool clSDLWindow::HandleEvent( const SDL_Event& Event )
{
  switch ( Event.type )
  {
    case SDL_WINDOWEVENT:
      if ( Event.window.event == SDL_WINDOWEVENT_SIZE_CHANGED)
      {
        m_Width  = static_cast<float>( Event.window.data1 );
        m_Height = static_cast<float>( Event.window.data2 );
      }
      return true;
    case SDL_KEYDOWN:
    case SDL_KEYUP:
      OnKey( Event.key.keysym.sym, Event.type == SDL_KEYDOWN );
      break;
    case SDL_MOUSEBUTTONDOWN:
    case SDL_MOUSEBUTTONUP:
      break;
    case SDL_MOUSEMOTION:
      break;
    case SDL_MOUSEWHEEL:
      break;
  }
  return true;
}
```

Not all of the case labels are implemented and not all SDL2 events are used. We will make use of this routing on an as-needed basis in the subsequent chapters.

In our example, we render a rotating box using some useful OpenGL wrappers, which can hide the differences between mobile and desktop versions of OpenGL. Here is the code of `OnStart()` printing the version of OpenGL into the system log and initializing vertex buffer objects and shader programs:

```
clPtr<clVertexAttribs> g_Box;
clPtr<clGLVertexArray> g_BoxVA;
clPtr<clGLSLShaderProgram> g_ShaderProgram;
void OnStart( const std::string& RootPath )
{
  LOGI( "Hello Android NDK!" );
  const char* GLVersion  =
    (const char*)LGL3->glGetString( GL_VERSION  );
  const char* GLVendor   =
    (const char*)LGL3->glGetString( GL_VENDOR   );
  const char* GLRenderer =
    (const char*)LGL3->glGetString( GL_RENDERER );
  LOGI( "GLVersion : %s\n", GLVersion );
  LOGI( "GLVendor  : %s\n", GLVendor  );
  LOGI( "GLRenderer: %s\n", GLRenderer );
```

First, we create an API-agnostic representation of a box mesh:

```
g_Box = clGeomServ::CreateAxisAlignedBox(
  LVector3(-1), LVector3(+1) );
```

Then, we feed it into OpenGL to create a vertex array using vertex buffer object:

```
g_BoxVA = make_intrusive<clGLVertexArray>();
g_BoxVA->SetVertexAttribs( g_Box );
```

The shader program is constructed from two string variables containing the source code of vertex and fragment shaders:

```
g_ShaderProgram = make_intrusive<clGLSLShaderProgram>(
  g_vShaderStr, g_fShaderStr );
LGL3->glClearColor( 0.1f, 0.0f, 0.0f, 1.0f );
LGL3->glEnable( GL_DEPTH_TEST );
}
```

Here are the shaders written in GLSL 3.3 Core Profile. Transform the vertices using the model-view-projection matrix:

```
static const char g_vShaderStr[] =
R"(
  uniform mat4 in_ModelViewProjectionMatrix;
```

```
    in vec4 in_Vertex;
    in vec2 in_TexCoord;
    out vec2 Coords;
    void main()
    {
      Coords = in_TexCoord.xy;
      gl_Position = in_ModelViewProjectionMatrix * in_Vertex;
    }
)";
```

Paint the box using texture coordinates as RG color components:

```
    static const char g_fShaderStr[] =
    R"(
      in vec2 Coords;
      out vec4 out_FragColor;
      void main()
      {
        out_FragColor = vec4( Coords, 1.0, 1.0 );
      }
)";
```

You may have noticed the source code of the shaders does not contain #version and precision lines. This is because the clGLSLShaderProgram class does some manipulations on the source code to abstract the differences between different versions of GLSL. We will familiarize ourselves with this class in the subsequent paragraphs. Before that, let's take a look at OnDrawFrame():

```
    void OnDrawFrame()
    {
      static float Angle = 0;
      Angle += 0.02f;
      LGL3->glClear( GL_COLOR_BUFFER_BIT | GL_DEPTH_BUFFER_BIT );
      mat4 Proj = Math::Perspective(
        45.0f, g_Window->GetAspect(), 0.4f, 2000.0f );
```

Rotate the cube around the (1, 1, 1) axis:

```
      LMatrix4 MV = LMatrix4::GetRotateMatrixAxis( Angle,
        vec3( 1, 1, 1 ) ) *
        mat4::GetTranslateMatrix( vec3( 0, 0, -5 ) );
      g_ShaderProgram->Bind();
      g_ShaderProgram->SetUniformNameMat4Array(
        "in_ModelViewProjectionMatrix", 1, MV * Proj );
      g_BoxVA->Draw( false );
    }
```

OpenGL API binding

As you can see, all OpenGL calls in the earlier mentioned code are done via the LGL3 prefix. This is a structure called sLGLAPI declared in LGLAPI.h containing pointers to actual OpenGL API functions:

```
struct sLGLAPI
{
  sLGLAPI()
  {
    memset( this, 0, sizeof( *this ) );
  };
  PFNGLACTIVETEXTUREPROC          glActiveTexture;
  PFNGLATTACHSHADERPROC           glAttachShader;
  PFNGLBINDATTRIBLOCATIONPROC     glBindAttribLocation;
  PFNGLBINDBUFFERPROC             glBindBuffer;
  PFNGLBINDBUFFERBASEPROC         glBindBufferBase;
  PFNGLBINDFRAGDATALOCATIONPROC glBindFragDataLocation;
  ...
}
```

The fields of the sLGLAPI structure are set in the LGL::GetAPI() function. There are two distinct implementations of this function, one is for Windows in LGL_Windows.h, and the other is in LGL_Android.h for everything else. The difference is in the dynamic linking on Windows, as shown in the following code:

```
void LGL::GetAPI( sLGLAPI* API )
{
  API->glBlendFunc =
    ( PFNGLBLENDFUNCPROC )GetGLProc( API, "glBlendFunc" );
  API->glBufferData =
    ( PFNGLBUFFERDATAPROC )GetGLProc( API, "glBufferData" );
  API->glBufferSubData =
    ( PFNGLBUFFERSUBDATAPROC )GetGLProc( API, "glBufferSubData");
  ...
}
```

All other platforms use static linking against the system-provided OpenGL library:

```
void LGL::GetAPI( sLGLAPI* API )
{
  API->glActiveTexture = &glActiveTexture;
  API->glAttachShader = &glAttachShader;
  API->glBindAttribLocation = &glBindAttribLocation;
  API->glBindBuffer = &glBindBuffer;
  ...
}
```

Of course, if you use vendor-specific OpenGL extensions, you can access them using dynamic linking and `glGetProcAddresss()` on any platform and that is where the `sLGLAPI` structure comes in handy:

This was the lowest of our abstraction levels on top of OpenGL. One might say this so -called layer does nothing. This is not true. Just take a look how a pointer to `glClearDepth()` is retrieved on Android. Instead of a direct function call, there is a stub for some reason:

```
API->glClearDepth = &Emulate_glClearDepth;
```

The stub is defined in the following way:

```
LGL_CALL void Emulate_glClearDepth( double Depth )
{
  glClearDepthf( static_cast<float>( Depth ) );
}
```

The reason is there was no `glClearDepth()` function in OpenGL ES, which accepts a `float` parameter but OpenGL 3 has one. This way the API difference between mobile and desktop OpenGL can be hidden from the client code behind a thin abstraction layer. Using this technique, you can transparently replace one OpenGL enums with the other. A tracing mechanism can be transparently implemented which can print the values of OpenGL function parameters into the log. This technique is crucial when porting existing applications to platforms where no graphics debugger is available (yes, we are looking at you, BlackBerry). We will leave that as an exercise for you.

Let's now dive deeper and find out how higher-level abstractions are implemented.

Cross-platform OpenGL abstractions

Geometric objects can be represented by their surfaces. In this chapter, we are talking only about polygonal graphics, so the data structure of fundamental importance is the *triangular mesh*.

Just as with the digital audio, our convenient API-agnostic data structures should be converted to something native for the graphics API before they can be rendered. Let's start with the representation of triangulated geometry in the 3D space.

A single triangle can be specified by three vertices. Each vertex stores at least its position in the 3D space, as shown in the following figure:

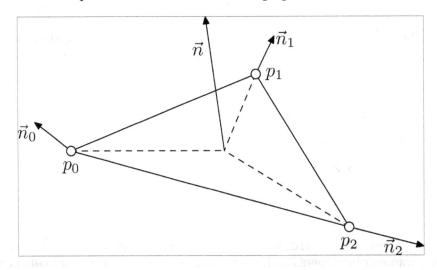

The first thing in implementing a portable renderer, we need to separate geometry storage, which in the most simple case, is just a collection of vertices with their attributes and the order to iterate through these vertices to construct graphical primitives, from any API-specific functions and data types. This kind of data structure is implemented in the `clVertexAttribs` class:

```
class clVertexAttribs: public iIntrusiveCounter
{
public:
  clVertexAttribs();
  explicit clVertexAttribs( size_t Vertices );
  void SetActiveVertexCount( size_t Count )
  {FActiveVertexCount = Count; }
  size_t GetActiveVertexCount() const
  { return FActiveVertexCount; }
```

This method returns a container of pointer to the actual vertex attributes, positions, texture coordinates, normals, and colors, which can be fed into an OpenGL vertex buffer object:

```
const std::vector<const void*>&
  EnumerateVertexStreams() const;
{
  FStreams[ L_VS_VERTEX   ] = &FVertices[0];
```

```
    FStreams[ L_VS_TEXCOORD ] = &FTexCoords[0];
    FStreams[ L_VS_NORMAL   ] = &FNormals[0];
    FStreams[ L_VS_COLORS   ] = &FColors[0];
    return FStreams;
}
```

We declare a bunch of helper methods to generate geometry data:

```
void Restart( size_t ReserveVertices );
void EmitVertexV( const vec3& Vec );
void SetTexCoordV( const vec2& V );
void SetNormalV( const vec3& Vec );
void SetColorV( const vec4& Vec );
```

We declare a set of public fields to store our data. The vertex 3D positions x, y, z are declared as follows:

```
public:
    std::vector<vec3> FVertices;
```

Texture coordinates u and v. This is a limitation of our vertex format since sometimes texture coordinates can contain more than two channels. However, for our applications, this limitation is appropriate and viable:

```
    std::vector<vec2> FTexCoords;
```

The vertex normals are usually in the object space:

```
    std::vector<vec3> FNormals;
```

RGBA colors of the vertices. This container can be used for any custom data you want if you write the right shader, of course:

```
    std::vector<vec4> FColors;
};
```

The implementation is simple; however, we suggest that you take a look at the Geometry.cpp and Geometry.h files before proceeding further.

To populate instances of clVertexAttribs with useful data, a set of static methods is declared within the clGeomServ class:

```
classlGeomServ
{
public:
```

```
static clPtr<clVertexAttribs> CreateTriangle2D(
  float vX, float vY, float dX, float dY, float Z );
static clPtr<clVertexAttribs> CreateRect2D(
  float X1, float Y1, float X2, float Y2, float Z,
  bool FlipTexCoordsVertical, int Subdivide );
static void AddAxisAlignedBox(
  const clPtr<clVertexAttribs>& VA,
  const LVector3& Min, const LVector3& Max );
static clPtr<clVertexAttribs> CreateAxisAlignedBox(
  const LVector3& Min, const LVector3& Max );
static void AddPlane( const clPtr<clVertexAttribs>& VA,
  float SizeX, float SizeY,
  int SegmentsX, int SegmentsY, float Z );
static clPtr<clVertexAttribs> CreatePlane(
  float SizeX, float SizeY, int SegmentsX, int SegmentsY,
  float Z );
};
```

All `Create*()` methods create a new geometry primitive and return an instance of `clVertexAttribs` containing it. Methods starting with `Add` add a primitive to the existing instance of the `clVertexAttribs` class assuming it has enough capacity to store the new primitive.Implementations of these mempleare trivial and can be found in `Geometry.cpp`. More sophisticated geometry generation routines will be added in the subsequent chapters.

Feeding the geometry data to OpenGL

To render the contents of `clVertexAttribs`, we need to convert its data into a set of API-specific buffers and API function invocations. This is done in the `clGLVertexArray` class by creating **Vertex Array Object (VOA)** and **Vertex Buffer Object (VBO)** OpenGL objects and fetching contents from `clVertexAttribs`:

```
class clGLVertexArray: public iInusiveCounter
{
public:
  clGLVertexArray();
  virtual ~clGLVertexArray();
```

The `Draw()` method does the actual rendering and it is the lowest level possible with our abstraction layer to actually render anything:

```
void Draw( bool Wireframe ) const;
void SetVertexAttribs(
  const clPtr<clVertexAttrs>& Attribs );
```

```
private:
  void Bind() const;
private:
  Luint FVBOID;
  Luint FVAOID;
```

These pointers are actually offsets of vertex data inside the vertex buffer:

```
std::vector<const void*> FAttribVBOOffset;
```

And these pointers point to the actual data from `clVertexAttribs`:

```
std::vector<const void*> FEnumeratedStreams;
clPtr<clVertexAttribs> FAttribs;
};
```

The implementation of this class includes some book-keeping and calling the OpenGL functions. The constructor and destructor initialize and destroy handles for VOA and VBO:

```
clGLVertexArray::clGLVertexArray()
: FVBOID( 0 ),
  FVAOID( 0 ),
  FAttribVBOOffset( L_VS_TOTAL_ATTRIBS ),
  FEnumeratedStreams( L_VS_TOTAL_ATTRIBS ),
  FAttribs( nullptr )
{
```

On Windows, we use OpenGL 4 where usage of vertex array object is mandatory:

```
#if defined( _WIN32 )
  LGL3->glGenVertexArrays( 1, &FVAOID );
#endif
}
```

Destruction is done in a similar platform-specific way:

```
clGLVertexArray::~clGLVertexArray()
{
  LGL3->glDeleteBuffers( 1, &FVBOID );
  #if defined( _WIN32 )
    LGL3->glDeleteVertexArrays( 1, &FVAOID );
  #endif
}
```

The private method `Bind()` sets this vertex array object as the source vertex stream for the OpenGL rendering pipeline:

```
void clGLVertexArray::Bind() const
{
  LGL3->glBindBuffer( GL_ARRAY_BUFFER, FVBOID );
  LGL3->glVertexAttribPointer( L_VS_VERTEX,
    L_VS_VEC_COMPONENTS[ 0 ], GL_FLOAT, GL_FALSE, 0,
    FAttribVBOOffset[ 0 ] );
  LGL3->glEnableVertexAttribArray( L_VS_VERTEX );
```

After binding and enabling the vertex positions, we enable each additional non-empty attribute:

```
for ( int i = 1; i < L_VS_TOTAL_ATTRIBS; i++ )
{
  LGL3->glVertexAttribPointer( i,
    L_VS_VEC_COMPONENTS[ i ],
    GL_FLOAT, GL_FALSE, 0, FAttribVBOOffset[ i ] );
  FAttribVBOOffset[ i ] ?
    LGL3->glEnableVertexAttribArray( i ) :
    LGL3->glDisableVertexAttribArray( i );
}
}
```

The `Draw()` method binds the VOA and calls `glDrawArrays()` to render the geometry:

```
void clGLVertexArray::Draw( bool Wireframe ) const
{
  #if defined( _WIN32 )
    LGL3->glBindVertexArray( FVAOID );
  #else
    Bind();
  #endif
```

The first parameter is the type of primitives. If the `Wireframe` parameter is `true`, we tell OpenGL to treat the data as a collection of lines, one for each sequential pair of points. If the parameter is `false`, each sequential point triple is used as three vertices of a triangle:

```
  LGL3->glDrawArrays(
    Wireframe ? GL_LINE_LOOP : GL_TRIANGLES, 0,
    static_cast<GLsizei>( FAttribs->GetActiveVertexCount() ) );
}
```

The `SetVertexAttribs()` member function attaches the geometry to `GLVertexArray` and recreates all the required OpenGL objects:

```
void clGLVertexArray::SetVertexAttribs(
  const clPtr<clVertexAttribs>& Attribs )
{
  FAttribs = Attribs;
```

After assigning a pointer, we acquire an array of pointers to individual vertex attribute streams:

```
FEnumeratedStreams = FAttribs->EnumerateVertexStreams();
LGL3->glDeleteBuffers( 1, &FVBOID );
size_t VertexCount = FAttribs->FVertices.size();
size_t DataSize = 0;
```

Every stream is checked if it contains any data and the size of the vertex buffer is updated accordingly:

```
for ( int i = 0; i != L_VS_TOTAL_ATTRIBS; i++ )
{
  FAttribVBOOffset[ i ] = ( void* )DataSize;
  DataSize += FEnumeratedStreams[i] ?
    sizeof( float ) * L_VS_VEC_COMPONENTS[ i ] *
    VertexCount : 0;
}
```

After this, we create a new vertex buffer object that will contain the geometry data:

```
LGL3->glGenBuffers( 1, &FVBOID );
LGL3->glBindBuffer( GL_ARRAY_BUFFER, FVBOID );
```

The most important thing here is to copy the data from `clVertexAttribs` object to GPU memory. This is done by calling `glBufferData()` with a `nullptr` value as the buffer pointer to allocate the storage:

```
LGL3->glBufferData(
  GL_ARRAY_BUFFER, DataSize, nullptr, GL_STREAM_DRAW );
```

You can find more information about `glBufferData()` at https://www.khronos.org/opengles/sdk/docs/man3/html/glBufferData.xhtml.

Here are subsequent calls to `glBufferSubData()` for each non-empty attribute array, those are vertex positions, texture coordinates, normals, and colors:

```
for ( int i = 0; i != L_VS_TOTAL_ATTRIBS; i++ )
{
  if ( FEnumeratedStreams[i] )
  {
    LGL3->glBufferSubData( GL_ARRAY_BUFFER,
      ( GLintptr )FAttribVBOOffset[ i ],
      FAttribs->GetActiveVertexCount() *
      sizeof( float ) * L_VS_VEC_COMPONENTS[ i ],
      FEnumeratedStreams[ i ] );
  }
}
```

Binding is somewhat specific for VAO and non-VAO versions:

```
#if defined( _WIN32 )
  LGL3->glBindVertexArray( FVAOID );
  Bind();
  LGL3->glBindVertexArray( 0 );
#endif
}
```

The VAO version can be used on OpenGL ES 3. However, unmodified code also runs on OpenGL ES 2.

Shader programs

Both desktop and mobile OpenGL versions use shader programs as parts of their rendering pipelines. Feeding just the geometry is not enough. However, there are several important differences between GLSL 3.00 ES and GLSL 3.30 Core we should deal with to create a portable rendering subsystem.

Let's start with the declaration of a `uniform` value:

```
struct sUniform
{
public:
  explicit sUniform( const std::string& e )
  : FName( e )
  , FLocation( -1 )
  {};
```

```
   sUniform( int Location, const std::string& e )
   : FName( e )
   , FLocation( Location )
   {};
   std::string FName;
   Lint FLocation;
};
```

This class stores name and location of a uniform within a linked shader program.
The shader program class looks as follows:

```
class clGLSLShaderProgram: public iIntrusiveCounr
{
public:
```

The constructor takes the source code of vertex and fragment shaders as parameters:

```
clGLSLShaderProgram( const std::string& VShader,
   const std::string& FShader );
virtual ~clGLSLShaderProgram();
```

The `Bind()` method binds the shader program before usage:

```
void Bind();
```

A group of methods dealing with uniforms:

```
Lint CreateUniform( const std::string& Name );
void SetUniformNameFloat( const std::string& Name,
   const float Float );
void SetUniformNameFloatArray( const std::string& Name,
   int Count, const float& Float );
void SetUniformNameVec3Array( const std::string& Name,
   int Count, const LVector3& Vector );
void SetUniformNameVec4Array( const std::string& Name,
   int Count, const LVector4& Vector );
void SetUniformNameMat4Array( const std::string& Name,
   int Count, const LMatrix4& Matr );
private:
```

Link the program using the attached shaders:

```
bool RelinkShaderProgram();
```

We need to bind default locations of attributes and fragment data. This is done in the following method:

```
void BindDefaultLocations( Luint ProgramID )
{
  LGL3->glBindAttribLocation( ProgramID, L_VS_VERTEX,
    "in_Vertex" );
  LGL3->glBindAttribLocation( ProgramID, L_VS_TEXCOORD,
    "in_TexCoord" );
  LGL3->glBindAttribLocation( ProgramID, L_VS_NORMAL,
    "in_Normal" );
  LGL3->glBindAttribLocation( ProgramID, L_VS_COLORS,
    "in_Color" );
  LGL3->glBindFragDataLocation( ProgramID, 0,
    "out_FragColor" );
}
```

It binds the shader variables in_Vertex, in_Normal, in_TexCoord, and in_Color to appropriate vertex streams. You can declare and use these in variables in your GLSL code. The out_FragColor output variable is associated with the single output of a fragment shader.

Compile and attach a shader to this shader program:

```
Luint AttachShaderID( Luint Target,
  const std::string& ShaderCode, Luint OldShaderID );
```

Check and log any errors occurred while compiling and linking:

```
bool CheckStatus( Luint ObjectID, Lenum Target,
  const std::string& Message ) const;
```

This method retrieves all uniforms from the linked shader program and stores them as sUniform structures within the FUniforms container:

```
void RebindAllUniforms();
private:
  std::string FVertexShader;
  std::string FFragmentShader;
  Luint FVertexShaderID;
  Luint FFragmentShaderID;
```

A collection of active uniforms in this shader program is stored as follows:

```
std::vector<sUniform> FUniforms;
```

An OpenGL shader program and shader identifiers are stored in the following fields:

```
    Luint FProgramID;
    std::vector<Luint> FShaderID;
};
clGLSLShaderProgram::clGLSLShaderProgram(
    const std::string& VShader, const std::string& FShader )
: FVertexShader( VShader )
, FFragmentShader( FShader )
, FUniforms()
, FProgramID( 0 )
, FVertexShaderID( 0 )
, FFragmentShaderID( 0 )
{
    RelinkShaderProgram();
}
```

We can destroy all the created OpenGL objects as follows:

```
clGLSLShaderProgram::~clGLSLShaderProgram()
{
    LGL3->glDeleteProgram( FProgramID );
    LGL3->glDeleteShader( FVertexShaderID );
    LGL3->glDeleteShader( FFragmentShaderID );
}
```

Let's see how a shader object is created and attached to the shader program:

```
Luint clGLSLShaderProgram::AttachShaderID( Luint Target,
    const std::string& ShaderCode, Luint OldShaderID )
{
```

Since we use OpenGL ES 3 and OpenGL 4, the version of shaders should be specified accordingly:

```
    #if defined( USE_OPENGL_4 )
        std::string ShaderStr = "#version 330 core\n";
    #else
        std::string ShaderStr = "#version 300 es\n";
        ShaderStr += "precision highp float;\n";
        ShaderStr += "#define USE_OPENGL_ES_3\n";
    #endif
        ShaderStr += ShaderCode;
```

The resulting shader is submitted to OpenGL API functions:

```
Luint Shader = LGL3->glCreateShader( Target );
const char* Code = ShaderStr.c_str();
LGL3->glShaderSource( Shader, 1, &Code, nullptr );
LGL3->glCompileShader( Shader );
```

Check the compilation status and log any errors detected while compiling the code. This code falls back to the previously compiled shader if the new one fails to compile. You can implement dynamic shader program reloading as an exercise using filesystem classes from the previous chapters:

```
if ( !CheckStatus( Shader, GL_COMPILE_STATUS,
    "Shader wasn''t compiled:" ) )
{
  LGL3->glDeleteShader( Shader );
  return OldShaderID;
}
if ( OldShaderID )
{
  LGL3->glDeleteShader( OldShaderID );
}
return Shader;
}
```

Error checking and logging is not that complicated to implement and is a must-have:

```
bool clGLSLShaderProgram::CheckStatus( Luint ObjectID,
  Lenum Target, const std::string& Message ) const
{
  Lint   SuccessFlag = 0;
  Lsizei Length      = 0;
  Lsizei MaxLength    = 0;
  if ( LGL3->glIsProgram( ObjectID ) )
  {
    LGL3->glGetProgramiv( ObjectID, Target, &SuccessFlag );
    LGL3->glGetProgramiv( ObjectID, GL_INFO_LOG_LENGTH,
      &MaxLength );
```

A buffer for a shader program error message is allocated dynamically on the stack:

```
char* Log = ( char* )alloca( MaxLength );
LGL3->glGetProgramInfoLog( ObjectID, MaxLength,
  &Length, Log );
if ( *Log ) { LOGI( "Program info:\n%s\n", Log ); }
}
```

```
      else if ( LGL3->glIsShader( ObjectID ) )
      {
        LGL3->glGetShaderiv( ObjectID, Target, &SuccessFlag );
        LGL3->glGetShaderiv( ObjectID, GL_INFO_LOG_LENGTH,
          &MaxLength );
```

Deal with a shader object in a similar way as follows:

```
        char* Log = ( char* )alloca( MaxLength );
        LGL3->glGetShaderInfoLog( ObjectID, MaxLength,
          &Length, Log );
        if ( *Log ) { LOGI( "Shader info:\n%s\n", Log ); }
      }
      return SuccessFlag != 0;
    }
```

Relinking of the shader program is done when both vertex and fragment shader objects had been successfully compiled:

```
    bool clGLSLShaderProgram::RelinkShaderProgram()
    {
      Luint ProgramID = LGL3->glCreateProgram();
      FVertexShaderID = AttachSaderID(
        GL_VERTEX_SHADER, FVertexShader, FVertexShaderID );
      if ( FVertexShaderID )
      { LGL3->glAttachShader( ProgramID, FVertexShaderID ); }
      FFragmentShaderID = AttachShaderID( GL_FRAGMENT_SHADER,
        FFragmentShader, FFragmentShaderID );
      if ( FFragmentShaderID )
      { LGL3->glAttachShader( ProgramID, FFragmentShaderID ); }
```

Bind locations of all default vertex attributes:

```
      BindDefaultLocations( ProgramID );
      LGL3->glLinkProgram( ProgramID );
      if ( !CheckStatus(
        ProgramID, GL_LINK_STATUS, "Program wasn''t linked" ) )
      {
        LOGI( "INTERNAL ERROR: Error while shader relinking" );
        return false;
      }
```

At this point, we know the shader program was linked successfully, and we can use it as a part of our rendering pipeline. Replace the old program with this code:

```
LGL3->glDeleteProgram( FProgramID );
FProgramID = ProgramID;
```

Retrieve the list of active uniforms from the linked program and store them:

```
RebindAllUniforms();
```

Bind texture samplers to their default locations. You can add more texture units whenever you desire:

```
LGL3->glUniform1i(
    LGL3->glGetUniformLocation(FProgramID, "Texture0"), 0);
LGL3->glUniform1i(
    LGL3->glGetUniformLocation(FProgramID, "Texture1"), 1);
LGL3->glUniform1i(
    LGL3->glGetUniformLocation(FProgramID, "Texture2"), 2);
LGL3->glUniform1i(
    LGL3->glGetUniformLocation(FProgramID, "Texture3"), 3);
return true;
}
```

Queuing of uniforms is done in the `RebindAllUniforms()` method:

```
void clGLSLShaderProgram::RebindAllUniforms()
{
  Bind();
  FUniforms.clear();
  Lint ActiveUniforms;
  char Buff[256];
  LGL3->glGetProgramiv( FProgramID,
    GL_ACTIVE_UNIFORMS, &ActiveUniforms );
  for ( int i = 0; i != ActiveUniforms; ++i )
  {
    Lsizei Length;
    Lint Size;
    Lenum Type;
    LGL3->glGetActiveUniform( FProgramID, i,
      sizeof( Buff ), &Length, &Size, &Type, Buff );
    std::string Name( Buff, Length );
```

The sUniform object is constructed and pushed into the container for future access. As an improvement, the vector can be sorted or replaced with std::map to allow faster access:

```
   sUniform Uniform( Name );
   Uniform.FLocation = LGL3->glGetUniformLocation(
      FProgramID, Name.c_str() );
   FUniforms.push_back( Uniform );
   }
}
```

The SetUniform*() group of methods sets the value for a named uniform in the GLSL shader program. These methods retrieve a handle of a uniform by calling CreateUniform(), and then use one of the glUniform*() OpenGL functions to set the new value. String names can be used for rapid prototyping of shaders. If you want to go for performance, retrieve the location of the uniform beforehand using the CreateUniform() member function and use that value with a corresponding call to SetUniform*():

```
   void clGLSLShaderProgram::SetUniformNameFloat(
      const std::string& Name, const float Float )
   {
      Lint Loc = CreateUniform( Name );
      LGL3->glUniform1f( Loc, Float );
   }
   void clGLSLShaderProgram::SetUniformNameFoatArray(
      const std::string& Name, int Count, const float& Float )
   {
      Lint Loc = CreateUniform( Name );
      LGL3->glUniform1fv( Loc, Count, &Float );
   }
```

Vectors are converted to pointers. Notice the following trick, the ToFloatPtr() method returns a pointer to the x component of a vector. In the case when this vector in packed into an array of vectors, we also have the pointer to the beginning of the array. Thus, the Count parameter makes perfect sense and we may pass arrays of vectors to this method:

```
   void clGLSLShaderProgram::SetUniformNameec3Array(
      const std::string& Name, int Count, const LVector3& Vector )
   {
      Lint Loc = CreateUniform( Name );
```

```
        LGL3->glUniform3fv( Loc, Count, Vector.ToFloatPtr() );
    }
    void clGLSLShaderProgram::SetUniformNameVec4Array(
      const std::string& Name, int Count, const LVector4& Vector )
    {
      Lint Loc = CreateUniform( Name );
      LGL3->glUniform4fv( Loc, Count, Vector.ToFloatPtr() );
    }
```

Methods for matrices differ from the previous ones only by parameter types:

```
    void clGLSLShaderProgram::SetUniformNameMat4Array(
      const std::string& Name, int Count, const LMatrix4& Matrix )
    {
      Lint Loc = CreateUniform( Name );
      LGL3->glUniformMatrix4fv( Loc, Count, false,
      Matrix.ToFloatPtr() );
    }
```

The `CreateUniform()` method used in `SetUniform*()` performs a search in the `FUniforms` container and returns the OpenGL identifier of the uniform:

```
    Lint clGLSLShaderProgram::CreateUniform(
      const std::string& Name )
    {
      for ( size_t i = 0; i != FUniforms.size(); ++i )
      if ( FUniforms[i].FName == Name )
      return FUniforms[i].FLocation;
      return -1;
    }
```

This method is safe to use for any name, since the value of -1 returned for uniforms not found in the shader program is accepted and ignored by OpenGL.

The `Bind()` method binds the shader program to the current OpenGL rendering context:

```
    void clGLSLShaderProgram::Bind()
    {
      LGL3->glUseProgram( FProgramID );
    }
```

In a more sophisticated application, it makes sense to cache the value of the currently binded shader program and call the underlying API only if the value was changed.

Textures

The last component we need to wrap is a texture. Textures are represented as instances of the clGLTexture class:

```
class clGLTexture: public iIntrusivounter
{
public:
   clGLTexture();
   virtual ~clGLTexture();
```

Bind the texture to a specified OpenGL texture unit:

```
void Bind( int TextureUnit ) const;
```

Load texture pixels from an API-independent bitmap:

```
void LoadFromBitmap( const clPtr<clBitmap>& Bitmap );
```

Set the texture coordinates clamping mode:

```
void SetClamping( Lenum Clamping );
```

Deal with the data formats and dimensions of t texture:

```
private:
   void SetFormat( Lenum Target, Lenum InternalFormat,
     Lenum Format, int Width, int Height );
   Luint FTexID;
   Lenum FInternalFormat;
   Lenum FFormat;
};
```

The implementation is quite compact. Here it is:

```
clGLTexturelGLTexture()
: FTexID( 0 )
, FIntelFormat( 0 )
, FFormat( 0 )
{
}
clGLTexture::~clGLTexture()
{
   if ( FTexID ) { LGL3->glDeleteTextures( 1, &FTexID ); }
}
void clGLTexture::Bind( int TextureUnit ) const
```

```
  {
    LGL3->glActiveTexture( GL_TEXTURE0 + TextureUnit );
    LGL3->glBindTexture( GL_TEXTURE_2D, FTexID );
  }
```

We can set the format of the texture without uploading any pixels. This is useful if you want to attach the texture to a frame buffer object. We will use this functionality in *Chapter 8, Writing a Rendering Engine,* to implement render-to-texture functionality:

```
void clGLTexture::SetFormat( Lenum Target, Lenum InternalFormat,
  Lenum Format, int Width, int Height )
{
  if ( FTexID )
  {
    LGL3->glDeleteTextures( 1, &FTexID );
  }
  LGL3->glGenTextures( 1, &FTexID );
  LGL3->glBindTexture( GL_TEXTURE_2D, FTexID );
  LGL3->glTexParameterf( GL_TEXTURE_2D,
    GL_TEXTURE_MIN_FILTER, GL_LINEAR );
  LGL3->glTexParameterf( GL_TEXTURE_2D,
    GL_TEXTURE_MAG_FILTER, GL_LINEAR );
  LGL3->glTexParameteri( GL_TEXTURE_2D,
    GL_TEXTURE_WRAP_S, GL_CLAMP_TO_EDGE );
  LGL3->glTexParameteri( GL_TEXTURE_2D,
    GL_TEXTURE_WRAP_T, GL_CLAMP_TO_EDGE );
  LGL3->glTexImage2D( GL_TEXTURE_2D, 0, InternalFormat,
    Width, Height, 0, Format, GL_UNSIGNED_BYTE, nullptr );
  LGL3->glBindTexture( GL_TEXTURE_2D, 0 );
}
void clGLTexture::SetClamping( Lenum Clamping )
{
  Bind( 0 );
```

Update S and T clamping modes as follows:

```
    LGL3->glTexParameteri( GL_TEXTURE_2D,
      GL_TEXTURE_WRAP_S, Clamping );
    LGL3->glTexParameteri( GL_TEXTURE_2D,
      GL_TEXTURE_WRAP_T, Clamping );
  }
void clGLTexture::LoadFromBitmap(
  const clPtr<clBitmap>& Bitmap )
{
  if ( !Bitmap ) { return; }
  if ( !FTexID )
  {
    LGL3->glGenTextures( 1, &FTexID );
  }
```

Choose an appropriate OpenGL texture format based on the bitmap parameters:

```
ChooseInternalFormat( Bitmap->FBitmapParams,
  &FFormat, &FInternalFormat );
Bind( 0 );
```

Set the default filtering mode to `GL_LINEAR` to avoid building a mipmap chain:

```
LGL3->glTexParameteri( GL_TEXTURE_2D,
  GL_TEXTURE_MIN_FILTER, GL_LINEAR );
LGL3->glTexParameteri( GL_TEXTURE_2D,
  GL_TEXTURE_MAG_FILTER, GL_LINEAR );
int Width = Bitmap->GetWidth();
int Height = Bitmap->GetHeight();
```

Some OpenGL ES implementations does not allow zero-size textures (yes, we are looking at you, Vivante):

```
if ( !Width || !Height ) { return; }
```

Load raw bitmap data into OpenGL:

```
LGL3->glTexImage2D( GL_TEXTURE_2D, 0, FInternalFormat,
  Width, Height, 0, FFormat, GL_UNSIGNED_BYTE,
  Bitmap->FBitmapData );
}
```

Up to this point, we have enough instruments at our disposal to build portable mobile applications using OpenGL. The example application `1_GLES` for this chapter renders a colored rotating cube on Windows and Android:

The Windows version can be compiled with `>make all -j16 -B`. An `.apk` package for Android can be built by calling these commands:

```
>ndk-build -j16 -B
>ant debug
```

Summary

We learned how to wrap raw OpenGL calls into a thin abstraction layer to hide many differences between OpenGL ES 3 and OpenGL 4. Now, let's proceed to the next chapter and learn how to implement basic graphical user interface rendering using OpenGL and classes shown in this chapter.

7
Cross-platform UI and Input System

In the previous chapter, we introduced classes and interfaces for platform-independent rendering. Here, we make a slight detour on our way to a 3D OpenGL renderer and use the SDL library to render elements of the user interface. To render our UI, we need lines, rectangles, textured rectangles, and text strings.

We will begin this chapter with the description of the iCanvas interface designed to render geometric primitives. The most complex part of iCanvas is Unicode text rendering, which is implemented using the FreeType library. Font glyphs caching is also a very important topic for complex UIs, which is discussed here. The second part of the chapter describes a multipage graphical user interface suitable for being the cornerstone for building interfaces of multiplatform applications. The chapter is concluded with an SDL application, which demonstrates capabilities of our UI system in action.

Rendering

Right now, we use only the SDL library without any OpenGL, so we will declare the iCanvas interface to allow immediate, but not always fast, rendering of geometric primitives and avoid creating the GLVertexArray instances described in the previous chapter. Later, we might provide a different iCanvas implementation to switch to another renderer:

```
class iCanvas: public iIntrusiveCounter
{
public:
```

The first two methods set the current rendering color specified as a triple of RGB integers or a 4-dimensional vector, which contains an additional alpha transparency value:

```
virtual void SetColor( int R, int G, int B ) = 0;
virtual void SetColor( const ivec4& C ) = 0;
```

The `Clear()` method clears the screen rendering surface:

```
virtual void Clear() = 0;
```

The `Rect()` and `Line()` methods render a rectangle and a line respectively, as their names suggest:

```
virtual void Rect( int X, int Y,
   int W, int H, bool Filled ) = 0;
virtual void Line( int X1, int Y1, int X2, int Y2 ) = 0;
```

The texture-related group of methods manages the creation and updation of textures. The `CreateTexture()` method returns an integer handle of the created texture. The texture handle `Idx` is passed as an argument to the `UpdateTexture()` member function to upload bitmap data into the texture. The `Pixels` parameter holds a bitmap object with pixel data:

```
virtual int CreateTexture( const clPtr<clBitmap>& Pixels ) = 0;
virtual int UpdateTexture( int Idx,
   const clPtr<clBitmap>& Pixels ) = 0;
virtual void DeleteTexture( int Idx ) = 0;
```

The `TextureRect()` method renders a quadrilateral using a specified texture:

```
virtual void TextureRect( int X, int Y, int W, int H,
   int SX, int SY, int SW, int SH, int Idx ) = 0;
```

The text rendering is done with a single `TextStr()` call, which specifies the rectangular area in which the text should be fit (or clamped), string to be rendered, font height in points, text color, and font ID from the `TextRenderer` class, which we will describe later:

```
virtual void TextStr( int X1, int Y1, int X2, int Y2,
   const std::string& Str, int Size,
   const LVector4i& Color, int FontID );
```

The last public member function is `Present()`, which ensures that all the primitives are shown on the screen:

```
virtual void Present() = 0;
};
```

We provide two implementations of the `iCanvas` interface. One uses the SDL library, the other is based on pure OpenGL calls. The `clSDLCanvas` class contains a pointer to an SDL renderer object `m_Renderer`. The constructor of `clSDLCanvas` takes a pointer to an instance of the `clSDLWindow` class, described in the previous chapter, to create a renderer attached to the window:

```
class clSDLCanvas: public iCanvas
{
private:
  SDL_Renderer* m_Renderer;
public:
  explicit clSDLCanvas( const clPtr<clSDLWindow>& Window )
  {
    m_Renderer = SDL_CreateRenderer(
      Window->GetSDLWindow(), -1, SDL_RENDERER_ACCELERATED );
  }
  virtual ~clSDLCanvas();
```

The `clSDLCanvas` class directly calls the corresponding SDL routines to render rectangles:

```
virtual void Rect( int X, int Y, int W, int H,
  bool Filled ) override
{
  SDL_Rect R = { X, Y, W, H };
  Filled ?
    SDL_RenderFillRect( m_Renderer, &R ) :
    SDL_RenderDrawRect( m_Renderer, &R );
}
```

The `SetColor()`, `Clear()`, and `Present()` member functions also call the appropriate SDL routines:

```
virtual void SetColor( int R, int G, int B ) override;
{
  SDL_SetRenderDrawColor( m_Renderer, R, G, B, 0xFF );
}
virtual void SetColor( const ivec4& C ) override;
{
  SDL_SetRenderDrawColor( m_Renderer, C.x, C.y, C.z, C.w );
}
virtual void Clear() override;
{
  SDL_RenderClear( m_Renderer );
}
```

```
virtual void Present() override
{
  SDL_RenderPresent( m_Renderer );
}
```

We have to do some bookkeeping to synchronize our clBitmap object with SDL_Texture. The internals look like this:

```
std::vector<SDL_Texture*> m_Textures;
```

The CreateTexture() method allocates a new SDL texture:

```
int CreateTexture( const clPtr<clBitmap>& Pixels )
{
  if ( !Pixels ) return -1;
  SDL_Texture* Tex = SDL_CreateTexture( m_Renderer,
    SDL_PIXELFORMAT_RGBA8888, SDL_TEXTUREACCESS_STREAMING,
    Pixels->GetWidth(), Pixels->GetHeight() );
  SDL_Rect Rect = { 0, 0, Pixels->GetWidth(),
    Pixels->GetHeight() };
```

We will use the pixel data within the Pixels object to update the SDL texture:

```
void* TexturePixels = nullptr;
int Pitch = 0;
int Result = SDL_LockTexture( Tex,
    &Rect, &TexturePixels, &Pitch );
```

Here, we assume the pitch of the texture is always equal to the pitch of our raw pixel data. This is not true in general. However, this assumption always holds for power-of-two textures. We suggest that you implement pitch-respecting texture updates as an exercise:

```
memcpy( TexturePixels, Pixels->FBitmapData,
    Pitch * Pixels->GetHeight() );
SDL_UnlockTexture(Tex);
```

After the texture is created, we store it in the m_Texture container:

```
int Idx = (int)m_Textures.size();
m_Textures.push_back( Tex );
return Idx;
}
```

The `UpdateTexture()` method is similar, except that it does not create a new texture and reuses the texture size from the previous one, hence, making updates much faster:

```
int UpdateTexture( int Idx, const clPtr<clBitmap>& Pixels )
{
  if ( !Pixels ) return;
  if ( !Pixels || Idx < 0 ||
    Idx >= (int)m_Textures.size() )
  {
    return -1;
  }
```

To update the texture, we will call the `SDL_LockTexture()` to get a pointer to the texture data and use `memcpy()` to copy bitmap pixels:

```
Uint32 Fmt;
int Access;
int W, H;
SDL_QueryTexture( m_Textures[Idx], &Fmt, &Access, &W, &H );
SDL_Rect Rect = { 0, 0, W, H };
void* TexturePixels = nullptr;
int Pitch = 0;
int res = SDL_LockTexture( m_Textures[Idx],
    &Rect, &TexturePixels, &Pitch );
```

Again, this only works for textures with the same pitch as in the provided bitmap:

```
memcpy( TexturePixels, Pixels->FBitmapData, Pitch * H );
SDL_UnlockTexture( m_Textures[Idx] );
}
```

When the texture is not required any more, it can be deleted using the `DeleteTexture()` member function:

```
void DeleteTexture( int Idx )
{
  if ( Idx < 0 || Idx >= (int)m_Textures.size() )
  {
    return;
  }
  SDL_DestroyTexture( m_Textures[Idx] );
  m_Textures[Idx] = 0;
}
```

The `TextureRect()` method calls the `SDL_RenderCopy()` function to draw a texture mapped rectangle:

```
void TextureRect( int X, int Y, int W, int H,
  int SX, int SY, int SW, int SH, int Idx )
{
  SDL_Rect DstRect = { X, Y, X + W, Y + H };
  SDL_Rect SrcRect = { SX, SY, SX + SW, SY + SH };
  SDL_RenderCopy( m_Renderer, m_Textures[Idx],
    &SrcRect, &DstRect);
}
```

The `TextStr()` method renders a UTF-8 encoded string into a rectangular region. It uses the FreeType library and requires some advanced machinery to work. We will discuss its implementation in the following sections. Let's have a look at the following:

```
virtual void TextStr(
    int X1, int Y1, int X2, int Y2,
    const std::string& Str, int Size,
    const LVector4i& Color, int FontID );
};
```

Basically, the `iCanvas` interface was designed around SDL, and its purpose is to hide the dependency on SDL behind a lightweight interface, so another implementation can be used relatively easily. Here, we implement the `iCanvas` interface using OpenGL and classes we introduced in the previous chapter. Take a look at the `clGLCanvas` class.

First, we need to define some GLSL shaders required to render filled and textured rectangles. We can do it naturally using the C++11 raw string literals. The vertex shader does remapping of window normalized coordinates, used in our canvas, into OpenGL normalized device coordinates and is shared between all fragment programs:

```
static const char RectvShaderStr[] =
R"(
  uniform vec4 u_RectSize;
  in vec4 in_Vertex;
  in vec2 in_TexCoord;
  out vec2 Coords;
  void main()
  {
    Coords = in_TexCoord;
```

```
        float X1 = u_RectSize.x;
        float Y1 = u_RectSize.y;
        float X2 = u_RectSize.z;
        float Y2 = u_RectSize.w;
        float Width = X2 - X1;
        float Height = Y2 - Y1;
```

We take the rectangle *0,0...1,1* and remap it into the required rectangle *X1,Y1-X2,Y2*. This way, we can use a single vertex array object to render any rectangle:

```
        vec4 VertexPos = vec4( X1 + in_Vertex.x * Width,
          Y1 + in_Vertex.y * Height, in_Vertex.z, in_Vertex.w ) *
          vec4( 2.0, -2.0, 1.0, 1.0 ) + vec4( -1.0, 1.0, 0.0, 0.0 );
        gl_Position = VertexPos;
      }
    )";
```

This fragment shader is used to render a flat colored rectangle:

```
    static const char RectfShaderStr[] =
    R"(
      uniform vec4 u_Color;
      out vec4 out_FragColor;
      in vec2 Coords;
      void main()
      {
        out_FragColor = u_Color;
      }
    )";
```

A texture-mapped version is marginally more complex. We modulate the constant color with a texture:

```
    static const char TexRectfShaderStr[] =
    R"(
      uniform vec4 u_Color;
      out vec4 out_FragColor;
      in vec2 Coords;
      uniform sampler2D Texture0;
      void main()
      {
        out_FragColor = u_Color * texture( Texture0, Coords );
      }
    )";
```

In the constructor of `clGLCanvas`, we will create all persistent OpenGL objects required for rendering:

```
clGLCanvas::clGLCanvas( const clPtr<clSDLWindow>& Window )
: m_Window( Window )
{
```

Initialize our OpenGL wrapper:

```
LGL3 = std::unique_ptr<sLGLAPI>( new sLGLAPI() );
LGL::GetAPI( LGL3.get() );
```

The geometry of this rectangle is reused to render rectangles of any dimensions:

```
m_Rect = clGeomServ::CreateRect2D(
  0.0f, 0.0f, 1.0f, 1.0f, 0.0f, false, 1 );
m_RectVA = new clGLVertexArray();
m_RectVA->SetVertexAttribs( m_Rect );
```

Link two shader programs from the source code:

```
m_RectSP = new clGLSLShaderProgram(
  RectvShaderStr, RectfShaderStr );
m_TexRectSP = new clGLSLShaderProgram(
  RectvShaderStr, TexRectfShaderStr );
}
```

A private helper function is used to convert integer window coordinates into normalized window coordinates that we use in our shaders:

```
vec4 clGLCanvas::ConvertScreenToNDC( int X, int Y, int W, int H )
const
{
  float WinW = static_cast<float>( m_Window->GetWidth() );
  float WinH = static_cast<float>( m_Window->GetHeight() );
  vec4 Pos( static_cast<float>( X ) / WinW,
    static_cast<float>( Y ) / WinH,
    static_cast<float>( X + W ) / WinW,
    static_cast<float>( Y + H ) / WinH );
  return Pos;
}
```

Now, the actual rendering code is very straightforward. Let's render a filled rectangle first:

```
void clGLCanvas::Rect( int X, int Y, int W, int H, bool Filled )
{
```

```
vec4 Pos = ConvertScreenToNDC( X, Y, W, H );
LGL3->glDisable( GL_DEPTH_TEST );
m_RectSP->Bind();
m_RectSP->SetUniformNameVec4Array( "u_Color", 1, m_Color );
m_RectSP->SetUniformNameVec4Array( "u_RectSize", 1, Pos );
```

Since alpha blending is a very costly operation, enable it only if the alpha channel of the color actually implies transparency:

```
if ( m_Color.w < 1.0f )
{
  LGL3->glBlendFunc(GL_SRC_ALPHA, GL_ONE_MINUS_SRC_ALPHA);
  LGL3->glEnable( GL_BLEND );
}
m_RectVA->Draw( false );
```

Disable blending again:

```
if ( m_Color.w < 1.0f )
{
  LGL3->glDisable( GL_BLEND );
}
}
```

Our implementation is very simple and does not do any state change tracking, which is very costly once you do a lot of Rect() calls. We would suggest that you add a method to the iCanvas interface, which can render a pack of rectangles at once sorting them into transparent and non-transparent buckets before rendering. This way multiple rectangles can be rendered reasonably fast. By the way, SDL does it in a similar fashion providing the SDL_FillRects() function.

Since we can use our clGLTexture class, texture management functions are now simple:

```
int clGLCanvas::CreateTexture( const clPtr<clBitmap>& Pixels )
{
  if ( !Pixels ) return -1;
  m_Textures.emplace_back( new clGLTexture() );
  m_Textures.back()->LoadFromBitmap( Pixels );
  return m_Textures.size()-1;
}
```

The `UpdateTexture()` and `DeleteTextures()` functions are almost one-liners, except the parameter validity check:

```
void clGLCanvas::UpdateTexture(
  int Idx, const clPtr<clBitmap>& Pixels )
{
  if ( m_Textures[ Idx ] ) m_Textures[ Idx ]->LoadFromBitmap(
    Pixels );
}
void clGLCanvas::DeleteTexture( int Idx )
{
  m_Textures[ Idx ] = nullptr;
}
```

Let's draw a textured rectangle using these textures. Most of the hassle is similar to `Rect()`, except the texture binding:

```
void clGLCanvas::TextureRect( int X, int Y, int W, int H,
  int SX, int SY, int SW, int SH, int Idx )
{
  if ( Idx < 0 || Idx >= (int)m_Textures.size() )
  {
    return;
  }
  vec4 Pos = ConvertScreenToNDC( X, Y, W, H );
  LGL3->glDisable( GL_DEPTH_TEST );
```

Bind the required texture to the texture unit 0:

```
  m_Textures[ Idx ]->Bind( 0 );
```

Use the `m_TexRectSP` shader program:

```
  m_TexRectSP->Bind();
  m_TexRectSP->SetUniformNameVec4Array( "u_Color", 1, m_Color );
  m_TexRectSP->SetUniformNameVec4Array( "u_RectSize", 1, Pos );
```

Blending is always used for textured rectangles since individual texels can be transparent:

```
  LGL3->glBlendFunc( GL_SRC_ALPHA, GL_ONE_MINUS_SRC_ALPHA );
  LGL3->glEnable( GL_BLEND );
  m_RectVA->Draw( false );
  LGL3->glDisable( GL_BLEND );
}
```

A similar optimization with the OpenGL state changes can be implemented here. We leave it to you to implement this caching mechanism. Now, let's proceed with the text rendering, so we can return to `clGLCanvas::TextStr()` afterwards.

Text rendering

In this section, we describe every essential detail of the text rendering process implemented in the `clTextRenderer` class. Here are the parts of our text renderer:

- UTF-8 string decoding (`http://en.wikipedia.org/wiki/UTF-8`)
- Text size calculation, kerning, and advance calculation
- Rendering of individual glyph, just like the one in the FreeType example from *Chapter 2*, *Native Libraries*
- Fonts and glyphs loading and caching
- String rendering

We assume all the strings are in the UTF-8 encoding because this way all Latin characters with ASCII codes between 0 and 127 take exactly one byte, and various national symbols take up to four bytes. The only problem with UTF-8 is that FreeType accepts fixed-width 2-byte UCS-2 encoding, so we have to include the decoding routine to convert from UTF-8 to UCS-2.

 There is a nice article on the absolute minimum every software developer must know about Unicode and character sets. Check it out at `http://www.joelonsoftware.com/articles/Unicode.html`.

We store each character of rendered string in the `FString` field of `clTextRenderer`:

```
class clTextRenderer
{
    std::vector<sFTChar> FString;
```

A character description is stored in the following structure with the UCS-2 character code in the `FChar` field and an internal character index `FIndex`:

```
struct sFTChar
{
    FT_UInt FChar;
    FT_UInt FIndex;
```

An `FGlyph` field holds the FreeType `FT_Glyph` structure with a rendered glyph:

```
FT_Glyph FGlyph;
```

After decoding character codes, we calculate the pixel width and advance for each glyph and store these values in `FWidth` and `FAdvance`:

```
FT_F26Dot6 FWidth;
FT_F26Dot6 FAdvance;
```

The `FCacheNode` field is used internally by FreeType font caching subsystem and is described briefly as follows:

```
FTC_Node FCacheNode;
```

The default constructor sets null values for each field:

```
sFTChar()
: FChar( 0 ), FIndex( ( FT_UInt )( -1 ) )
, FGlyph( nullptr ), FAdvance( 0 )
, FWidth( 0 ), FCacheNode( nullptr )
{ }
};
```

Now that we have a structure to hold our characters, we show how to process a string and calculate positions for each character. The subsequent paragraphs of this chapter describe the internal details of `clTextRenderer`, so that when we declare new fields, they are meant to be in the `clTextRenderer` class. We start with high-level routines that can render strings. After this, we get to UTF-8 decoding, and finally, show how to implement fonts management and caching.

Calculating glyph positions and string size in pixels

The `LoadStringWithFont()` member function takes a text string, an internal font identifier and desired font height in pixels. It calculates parameters of each element in the `FString` array. This routine is used in rendering and text size calculation:

```
bool TextRenderer::LoadStringWithFont(
  const std::string& S, int ID, int Height )
{
  if ( ID < 0 ) { return false; }
```

First, we get the font handle and determine if we need kerning. FFace is a field of type FT_Face within clTextRenderer. The GetSizedFace() method retrieves the font matching the desired height. It uses internal font cache to avoid rendering the game glyphs multiple times for a single resolution and is discussed in great detail later in this chapter. Take a look at the following code:

```
FFace = GetSizedFace( ID, Height );
if ( !FFace ) { return false; }
bool UseKerning = FT_HAS_KERNING( FFace );
```

Then, we decode the UTF-8 string to UCS-2 and fill the FString array:

```
DecodeUTF8( S.c_str() );
```

After FString is filled, we render each character and calculate positions:

```
for ( size_t i = 0, count = FString.size(); i != count; i++ )
{
  sFTChar& Char = FString[i];
  FT_UInt ch = Char.FChar;
```

First, we get the character index for the font and skip the end-of-line and carriage return characters:

```
Char.FIndex = ( ch != '\r' && ch != '\n' ) ?
  GetCharIndex( ID, ch ) : -1;
```

Once we know the index of the character, we may call the FT_RenderGlyph() method, but this is quite suboptimal to render a single glyph each time it is encountered. The GetGlyph() routine does all the work to extract a glyph from the cache:

```
Char.FGlyph = ( Char.FIndex != -1 ) ?
  GetGlyph( ID, Height, ch,
    FT_LOAD_RENDER, &Char.FCacheNode ) : nullptr;
```

If the glyph has been loaded successfully, we call the SetAdvance() method:

```
if ( !Char.FGlyph || Char.FIndex == -1 ) continue;
SetAdvance( Char );
```

Optionally, we can call the `Kern()` method to adjust the advance of the current character:

```
    if ( i > 0 && UseKerning )
    {
      Kern( FString[i - 1], Char );
    }
  }
  return true;
}
```

The auxiliary `SetAdvance()` method calculates the bounding box of a glyph and stores its width and advance in the `sFTChar` structure:

```
void TextRenderer::SetAdvance( sFTChar& Char )
{
  Char.FAdvance = Char.FWidth = 0;
  if ( !Char.FGlyph ) return;
```

The advance is stored as a $22:10$ fixed-point value, we convert it to an integer value using a bitwise shift:

```
  Char.FAdvance = Char.FGlyph->advance.x >> 10;
```

The `FT_Glyph_Get_CBox()` function returns a bounding box; we use its `xMax` field:

```
  FT_BBox bbox;
  FT_Glyph_Get_CBoxPTR( Char.FGlyph,
    FT_GLYPH_BBOX_GRIDFIT, &bbox );
  Char.FWidth = bbox.xMax;
```

For some glyphs, such as a whitespace, the width is zero and we use the `FAdvance` field:

```
  if ( Char.FWidth == 0 && Char.FAdvance != 0 )
  {
    Char.FWidth = Char.FAdvance;
  }
}
```

The `Kern()` routine takes two adjacent characters and calculates the advance correction. Our text renderer does not support automatic ligature substitution, and this might be the place to do it if such a substitution is required:

```
void TextRenderer::Kern( sFTChar& Left, const sFTChar& Right )
{
```

No kerning is required at the beginning and at the end of the string:

```
if ( Left.FIndex == -1 || Right.FIndex == -1 ) return;
FT_Vector Delta;
```

The FT_GetKerning() call calculates the relative offset correction for the current character:

```
FT_Get_KerningPTR( FFace, Left.FIndex, Right.FIndex,
  FT_KERNING_DEFAULT, &Delta );
```

The result is added to the advance value:

```
Left.FAdvance += Delta.x;
}
```

Using the FString array, we calculate the rendered string size easily by summing sizes of individual characters. Later, this size value is used to allocate the output bitmap for the string:

```
void TextRenderer::CalculateLineParameters( int* Width,
  int* MinY, int* MaxY, int* BaseLine ) const
{
```

The StrMinY and StrMaxY variables hold minimum and maximum pixel coordinates of a character in a string:

```
int StrMinY = -1000, StrMaxY = -1000;
if ( FString.empty() ) StrMinY = StrMaxY = 0;
```

The SizeX variable holds the number of horizontal pixels in the string bitmap. We iterate the FString array and add the advance of each character to SizeX:

```
int SizeX = 0;
for ( size_t i = 0 ; i != FString.size(); i++ )
{
    if ( FString[i].FGlyph == nullptr ) continue;
```

For each character, we get the glyph's bitmap and update the SizeX variable:

```
FT_BitmapGlyph BmpGlyph =
  ( FT_BitmapGlyph )FString[i].FGlyph;
SizeX += FString[i].FAdvance;
int Y = BmpGlyph->top;
int H = BmpGlyph->bitmap.rows;
```

After reading off the dimensions of a glyph, we update the minimum and maximum dimensions of the string:

```
        if ( Y > StrMinY ) StrMinY = Y;
        if ( H - Y > StrMaxY ) StrMaxY = H - Y;
    }
```

Finally, we calculate the integer value `Width` of the string by converting the $26:6$ fixed-point value `SizeX` to pixels:

```
    if ( Width ) { *Width = ( SizeX >> 6 ); }
    if ( BaseLine ) { *BaseLine = StrMaxY; }
    if ( MinY ) { *MinY = StrMinY; }
    if ( MaxY ) { *MaxY = StrMaxY; }
  }
```

Before doing glyphs rendering, we still should check out yet another important thing. Let's outline the process of UTF-8 characters decoding.

Decoding UTF-8

The `DecodeUTF8()` routine mentioned in the preceding section, which was used in `LoadStringWithFont()`, iterates the incoming byte array and uses `DecodeNextUTF8Char()` to get the character code in the UCS-2 encoding:

```
    bool TextRenderer::DecodeUTF8 ( const char* InStr )
    {
```

First, we store a pointer to a buffer and set the current position to zero:

```
    FIndex = 0;
    FBuffer = InStr;
```

The `FLength` field contains the number of bytes in `InStr`. The method `DecodeNextUTF8Char()` uses `FLength` to stop the decoding process when the end of string is reached:

```
    FLength = ( int )strlen( InStr );
    FString.clear();
    int R = DecodeNextUTF8Char();
```

Then, we will iterate the vector of bytes in `FBuffer` until the zero byte is encountered:

```
    while ( ( R != UTF8_LINE_END ) && ( R != UTF8_DECODE_ERROR ) )
    {
      sFTChar Ch;
```

The UCS-2 character code is the only thing we change in the new `sFTChar` instance:

```
      Ch.FChar = R;
      FString.push_back( Ch );
      R = DecodeNextUTF8Char();
   }
   return ( R != UTF8_DECODE_ERROR );
}
```

The UTF-8 decoder in `DecodeNextUTF8Char()` is based on the source code from the JSON checker, which can be downloaded from `http://www.json.org/ JSON_checker/utf8_decode.c`. To save space, we omit rather straightforward bit manipulation. The low-level implementation details can be found in the accompanying source code, just take a look at `TextRenderer.h` and `TextRenderer.cpp`.

Glyphs rendering

The `RenderLineOnBitmap()` method takes an allocated bitmap as an output surface and renders a given text string using a specified font identifier. The `LeftToRight` parameter tells us whether the text is written from left to right or from right to left:

```
   void TextRenderer:

   {
      LoadStringWithFont( TextString, FontID, FontHeight );
```

After loading, text rendering is done by iterating the `FString` container once again and calling the `DrawGlyphOnBitmap()` method for each character:

```
      int x = StartX << 6;
      for ( size_t j = 0 ; j != FString.size(); j++ )
      {
        if ( FString[j].FGlyph != 0 )
        {
          FT_BitmapGlyph BmpGlyph =
            ( FT_BitmapGlyph ) FString[j].FGlyph;
```

We track the current horizontal position in the x variable by summing advances of each character. For each non-empty glyph, we calculate a *real* onscreen position taking into account the actual text direction specified by the LeftToRight argument:

```
int in_x = ( x >> 6 ) +
   ( LeftToRight ? 1 : -1 ) * BmpGlyph->left;
```

If the direction is right to left, we will correct the position accordingly:

```
if ( !LeftToRight )
{
  in_x += BmpGlyph->bitmap.width;
  in_x = StartX + ( StartX - in_x );
}
DrawGlyphOnBitmap( Out, &BmpGlyph->bitmap,
  in_x, Y - BmpGlyph->top, Color );
}
```

At the end of each iteration, we shift the horizontal counter using the advance value:

```
  x += FString[j].FAdvance;
}
}
```

A wrapper routine RenderTextWithFont() precalculates the size of an output bitmap and returns a ready to use image:

```
clPtr<clBitmap> TextRenderer::RenderTextWithFont(
  const std::string& TextString, int FontID, int FontHeight,
  const ivec4& Color, bool LeftToRight )
{
  if ( !LoadStringWithFont( TextString, FontID, FontHeight ) )
  { return nullptr; }
  int W, Y;
  int MinY, MaxY;
  CalculateLineParameters( &W, &MinY, &MaxY, &Y );
  int H2 = MaxY + MinY;
```

After we have calculated the text size, we allocate an output bitmap, clear it, and call the RenderLineOnBitmap() method:

```
clPtr<clBitmap> Result =
  make_intrusive<clBitmap>( W, H2, L_BITMAP_BGRA8 );
Result->Clear();
```

The `RenderLineOnBitmap()` call fixes the starting position for the right-to-left text:

```
RenderLineOnBitmap(
  TextString, FontID, FontHeight, LeftToRight ? 0 : W - 1,
  MinY, Color, LeftToRight, Result );
return Result;
}
```

The `DrawGlyphOnBitmap()` method is similar to the code we used in *Chapter 2, Native Libraries*. We iterate through all pixels of the glyph's bitmap and set them using the data returned by FreeType:

```
void TextRenderer::DrawGlyphOnBitmap( const clPtr<clBitmap>& Out,
  FT_Bitmap* Bitmap, int X0, int Y0, const ivec4& Color ) const
{
  int W = Out->GetWidth();
  int Width = W - X0;
  if ( Width > Bitmap->width ) { Width = Bitmap->width; }
  for ( int Y = Y0 ; Y < Y0 + Bitmap->rows ; ++Y )
  {
    unsigned char* Src = Bitmap->buffer +
      ( Y - Y0 ) * Bitmap->pitch;
```

In the mask creation mode, we can copy the glyph directly into the output bitmap ignoring the `Color` parameter — that is, only a grayscale mask is rendered:

```
    if ( FMaskMode )
    {
      for ( int X = X0 + 0 ; X < X0 + Width ; X++ )
      {
        int Int = *Src++;
        int OutMaskCol = ( Int & 0xFF );
        Out->SetPixel(X, Y,
          ivec4i(OutMaskCol,
            OutMaskCol, OutMaskCol, 255) );
      }
    }
    else
```

For colored rendering, we fetch the source pixel and blend it with the specified color according to the mask:

```
    {
      for ( int X = X0 + 0 ; X < X0 + Width ; X++ )
      {
```

```
      unsigned int Int = *Src++;
      ivec4 Col = BlendColors(Color,
        Out->GetPixel(X, Y), (Int & 0xFF));
      if ( Int > 0 )
      {
        Col.w = Int;
        Out->SetPixel(X, Y, Col);
      }
    }
  }
}
```

The `BlendColors()` routine performs linear interpolation between color `C1` and `C2`. The right shifts here replace the division by 256. To avoid floating point arithmetic and conversions, the blend factor varies from 0 to 255, thus the value 255 instead of `1.0f` in the formula:

```
inline LVector4i BlendColors(
  const LVector4i& C1, const LVector4i& C2, unsigned int F )
{
  int r = ((C1.x) * F >> 8) + ((C2.x) * (255 - F) >> 8);
  int g = ((C1.y) * F >> 8) + ((C2.y) * (255 - F) >> 8);
  int b = ((C1.z) * F >> 8) + ((C2.z) * (255 - F) >> 8);
  return LVector4i(r, g, b, 255);
}
```

Now, we know how to render glyphs. Let's find out how to load, manage, and cache different fonts.

Font initialization and caching

Until now, we haven't described the details of font management, glyph rendering, and reusing characters' bitmaps.

First of all, we will declare a FreeType library handle used by every call to the FreeType API:

```
FT_Library FLibrary;
```

For each font we use, we need a rendered glyph cache and a character map cache. These caches are maintained by an `FTC_Manager` instance:

```
FTC_Manager FManager;
```

Next, we need glyph and character map caches:

```
FTC_ImageCache FImageCache;
FTC_CMapCache FCMapCache;
```

We keep track of byte buffers with loaded font files in the `FAllocatedFonts` field. The key for `std::map` is the name of a font file:

```
std::map<std::string, void*> FAllocatedFonts;
```

The `FFontFaceHandles` map is another container of initialized FreeType font handles:

```
std::map<std::string, FT_Face> FFontFaceHandles;
```

The private `LoadFontFile()` method reads a font file using our virtual filesystem mechanism and adds the initialized font to the containers declared in the preceding code:

```
FT_Error clTextRenderer::LoadFontFile(
  const std::string& FileName )
{
  if ( !FInitialized ) { return -1; }
```

We prevent reloading of already loaded fonts:

```
  if ( FAllocatedFonts.count( FileName ) > 0 ) { return 0; }
```

The new font is read into the `clBlob` object and its data is copied into a separate `Data` buffer:

```
clPtr<clBlob> DataBlob = LoadFileAsBlob(g_FS, FileName);
int DataSize = DataBlob->GetSize();
char* Data = new char[DataSize];
memcpy( Data, DataBlob->GetData(), DataSize );
```

The `FT_New_Memory_Face()` function creates a new `FT_Face` object, which is then stored in the `FFontFaceHandles` array:

```
FT_Face TheFace;
FT_Error Result = FT_New_Memory_FacePTR( FLibrary,
  ( FT_Byte* )Data, ( FT_Long )DataSize, 0, &TheFace );
if ( Result == 0 )
{
  FFontFaceHandles[ FileName ] = TheFace;
```

The `Data` buffer is added to `FAllocatedFonts` and the name of the font is added to the `FFontFaces` container:

```
      FAllocatedFonts[ FileName ] = ( void* )Data;
      FFontFaces.push_back( FileName );
   }
   return Result;
}
```

The `clTextRenderer` class we are developing contains the initialization code inside the `InitFreeType()` method:

```
   void clTextRenderer::InitFreeType()
   {
```

We omit the `LoadFT()` method description here because for Windows, it is a simple loading of a FreeType dynamic library file and resolution of function pointers. For Android, this method is empty and returns `true`:

```
   FInitialized = LoadFT();
   if ( FInitialized )
   {
      FInitialized = false;
```

The actual initialization code creates an instance of the FreeType library and allocates caches:

```
      if ( FT_Init_FreeTypePTR( &FLibrary ) != 0 ) { return; }
```

A cache manager is initialized after FreeType. The `FreeType_Face_Requester` is a function pointer to the method we describe in the following code. It resolves the font filename and does the actual loading of font data:

```
      if ( FTC_Manager_NewPTR(
         FLibrary, 0, 0, 0, FreeType_Face_Requester,
         this, &FManager ) != 0 )
      { return; }
```

At last, two caches are initialized similarly to the manager:

```
      if ( FTC_ImageCache_NewPTR( FManager,
         &FImageCache ) != 0)
      {
         return;
      }
```

```
     if ( FTC_CMapCache_NewPTR( FManager,
       &FCMapCache ) != 0 )
     {
       return;
     }
     FInitialized = true;
   }
 }
```

Deinitialization of FreeType is done in the reverse order:

```
   void TextRenderer::StopFreeType()
   {
```

First, we will clear the FString container by calling FreeString:

```
     FreeString();
```

Then, we will deallocate memory blocks with font data inside the FAllocatedFonts map:

```
     for ( auto p = FAllocatedFonts.begin();
       p != FAllocatedFonts.end() ; p++ )
     {
       delete[] ( char* )( p->second );
     }
```

At the end, we clear the container of font faces and destroy the cache manager and the library instance:

```
     FFontFaces.clear();
     if ( FManager ) { FTC_Manager_DonePTR( FManager ); }
     if ( FLibrary ) { FT_Done_FreeTypePTR( FLibrary ); }
   }
```

The FreeString method destroys the cached glyph for each element of the FString vector:

```
   void TextRenderer::FreeString()
   {
     for ( size_t i = 0 ; i < FString.size() ; i++ )
       if ( FString[i].FCacheNode != nullptr )
         FTC_Node_UnrefPTR( FString[i].FCacheNode,
           FManager );
     FString.clear();
   }
```

When FreeType finds out that there is no required font in the cache, it calls our `FreeType_Face_Requester()` callback to initialize a new font face:

```
FT_Error TextRenderer::FreeType_Face_Requester(
  FTC_FaceID FaceID,
  FT_Library Library,
  FT_Pointer RequestData,
  FT_Face* TheFace )
{
```

This is one of the awkward places where we really need to convert a C-style font face pointer into an integer identifier. We use lower 32 bits as an identifier:

```
#if defined(_WIN64) || defined(__x86_64__)
  long long int Idx = ( long long int )FaceID;
  int FaceIdx = ( int )( Idx & 0x7FFFFFFF );
#else
  int FaceIdx = reinterpret_cast< int >( FaceID );
#endif
```

If `FaceIdx` is less than zero, it is a valid pointer and the font was already loaded:

```
if ( FaceIdx < 0 ) { return 1; }
```

The method we are describing is a callback for a C language library, so we emulate the `this` pointer using `RequestData`. In the `InitFreeType()` method, we supplied `this` as a parameter to `FTC_Manager_New`:

```
clTextRenderer* This = ( clTextRenderer* )RequestData;
```

We extract a filename from the `FFontFaces` array:

```
std::string FileName = This->FFontFaces[FaceIdx];
```

The call to `LoadFontFile()` might return zero if we have already loaded the file:

```
FT_Error LoadResult = This->LoadFontFile( FileName );
```

If we haven't loaded the file, we search for the face in the `FFontFaceHandles` array:

```
*TheFace = ( LoadResult == 0 ) ?
  This->FFontFaceHandles[FileName] : nullptr;
return LoadResult;
}
```

We are getting close to the complete picture of `clTextRenderer` and only a few methods related to fonts and glyphs remain. The first one is `GetSizedFace()`, which we have used in `LoadStringWithFont()`:

```
FT_Face clTextRenderer::GetSizedFace( int FontID, int Height )
{
```

To start rendering glyphs at a given font height, we fill the `FTC_ScalerRec` structure to set rendering parameters. The `IntToID()` routine converts an integer identifier to a void pointer conversely to the code in `FreeType_Face_Requester()`:

```
FTC_ScalerRec Scaler;
Scaler.face_id = IntToID( FontID );
Scaler.height = Height;
Scaler.width = 0;
Scaler.pixel = 1;
FT_Size SizedFont;
```

The `FTC_Manager_LookupSize()` function searches for the `FT_Size` structure in the cache, which we supply to `FT_ActivateSize()`. After this, our glyphs get rendered with the size equal to the `Height` parameter:

```
if ( FTC_Manager_LookupSizePTR( FManager, &Scaler,
  &SizedFont ) != 0 ) return nullptr;
if ( FT_Activate_SizePTR( SizedFont ) != 0 ) return nullptr;
return SizedFont->face;
}
```

The second auxiliary method is `GetGlyph()`, which renders a single glyph:

```
FT_Glyph TextRenderer::GetGlyph(
  int FontID, int Height, FT_UInt Char,
  FT_UInt LoadFlags, FTC_Node* CNode )
{
```

Here, we convert a UCS-2 code to a character index:

```
FT_UInt Index = GetCharIndex( FontID, Char );
```

The `ImageType` structure is filled with glyph rendering parameters:

```
FTC_ImageTypeRec ImageType;
ImageType.face_id = IntToID( FontID );
ImageType.height = Height;
ImageType.width = 0;
ImageType.flags = LoadFlags;
```

Then, the `FTC_ImageCache_Lookup()` function searches for the previously rendered glyph and renders one if it has not been rendered yet:

```
    FT_Glyph Glyph;
    if ( FTC_ImageCache_LookupPTR( FImageCache,
      &ImageType, Index, &Glyph, CNode ) != 0 )
    { return nullptr; }
    return Glyph;
}
```

The third method `GetCharIndex()` uses the FreeType character map cache to quickly convert a UCS-2 character code to a glyph index:

```
    FT_UInt clTextRenderer::GetCharIndex( int FontID, FT_UInt Char )
    {
      return FTC_CMapCache_LookupPTR( FCMapCache,
        IntToID( FontID ), -1, Char );
    }
```

The `IntToID()` routine is similar to the casting code in `FreeType_Face_Requester()`. All it does is the conversion of an integer font face identifier to a C void pointer:

```
    inline void* IntToID( int FontID )
    {
      #if defined(_WIN64) || defined (__x86_64__)
        long long int Idx = FontID;
      #else
        int Idx = FontID;
      #endif
        FTC_FaceID ID = reinterpret_cast<void*>( Idx );
      return ID;
    }
```

Finally, we need the `GetFontHandle()` method, which loads a font file and returns a new valid font face identifier:

```
    int clTextRenderer::GetFontHandle( const std::string& FileName )
    {
```

First, we will try to load the file. The result of zero can be returned if the file is already loaded:

```
    if ( LoadFontFile( FileName ) != 0 )
    return -1;
```

We search for this font in a FFontFaces container and return its index:

```
for ( int i = 0 ; i != ( int )FFontFaces.size() ; i++ ) { }
if ( FFontFaces[i] == FileName )
   return i;
return -1;
}
```

We have all components necessary to render Unicode characters on bitmaps. Let's see how we can use this functionality to extend clCanvas with text rendering capabilities.

Integrating the text renderer into the canvas

Now that we have the clTextRenderer class, we can implement clGLCanvas::TextStr():

```
void clGLCanvas::TextStr( int X1, int Y1, int X2, int Y2,
   const std::string& Str, int Size, const ivec4& Color, int FontID )
{
```

First, we render the string to a bitmap:

```
auto B = TextRenderer::Instance()->RenderTextWithFont(
   Str, FontID, Size, Color, true );
```

A static texture is shared between all calls to TextStr(). Not that performant and multithreaded, but very simple:

```
static int Texture = this->CreateTexture();
```

Then, we update the static texture from this bitmap:

```
UpdateTexture( Texture, B );
```

After calculating the output size, we will call the TextureRect() method to render the bitmap with our text string:

```
int SW = X2 - X1 + 1, SH = Y2 - Y1 + 1;
this->TextureRect( X1, Y1, X2 - X1 + 1, Y2 - Y1 + 1,
   0, 0, SW, SH, Texture );
}
```

Global access to the single instance of `clTextRenderer` is implemented using the Singleton pattern:

```
clTextRenderer* clTextRenderer::Instance()
{
  static clTextRenderer Instance;
  return &Instance;
}
```

We can now render text using the `iCanvas` interface. Let's draw a graphical user interface where we can put our text.

Organizing the UI system

Having created the `iCanvas` interface for immediate mode rendering, we can switch to the user interface implementation. To create a meaningful application, the ability to render static or even animated graphical information is not always enough. An application must react to user input, which for mobiles often means responding to touch screen events. Here, we create a minimalistic graphical user interface consisting of three basic elements called views:

- `clUIView`: This is a logical container and a base class for other views
- `clUIStatic`: This is a static label with a text
- `clUIButton`: This is an object that fires events once it is touched

Each view is a rectangular region, which is capable of rendering itself and reacting to external events such as timing and user touches. Since we are working with the NDK, and at the same time, we want to debug our software on desktop machines, we must redirect events from an OS-specific queue to the C++ event handling code.

The base UI view

We define the `clUIView` interface for each UI element. This interface includes geometrical properties of the UI view:

```
class clUIView: public iIntrusiveCounter
{
protected:
```

This class contains geometric properties of the UI element. The m_X and m_Y fields contain relative coordinates in parent's coordinate frame. The m_ScreenX and m_ScreenY fields contain absolute coordinates in the screen reference frame. The m_Width and m_Height fields store the width and height of the element respectively:

```
int m_X, m_Y;
int m_ScreenX, m_ScreenY;
int m_Width, m_Height;
```

The private part of the class contains flags and settings for the child views layout. These settings are used later in the LayoutChildViews() method. The m_ParentFractionX and m_ParentFractionY values are used to override m_Width and m_Height as a percentage of the parent view size. If the values are greater than one, they are ignored. Their explicit usage is shown in the LayoutChildViews. The m_AlignV and m_AlignH contain different alignment modes for the coordinates:

```
private:
    float m_ParentFractionX, m_ParentFractionY;
    eAlignV m_AlignV;
    eAlignH m_AlignH;
    int m_FillMode;
```

The last field is the m_ChildViews vector with pointers to child views, as the name suggests:

```
std::vector< clPtr<clUIView> > m_ChildViews;
```

The default constructor sets the initial value for each field:

```
public:
    clUIView():
    m_X( 0 ), m_Y( 0 ), m_Width( 0 ), m_Height( 0 ),
    m_ScreenX( 0 ), m_ScreenY( 0 ), m_ParentFractionX( 1.0f ),
    m_ParentFractionY( 1.0f ), m_AlignV( eAlignV_DontCare ),
    m_AlignH( eAlignH_DontCare ), m_ChildViews( 0 )
    {}
```

The class interface contains Get* and Set* one-liners to access properties

```
virtual void SetPosition( int X, int Y ) { m_X = X; m_Y = Y; }
virtual void SetSize( int W, int H )
{ m_Width = W; m_Height = H; }
virtual void SetWidth( int W ) { m_Width = W; }
virtual void SetHeight( int H ) { m_Height = H; }
```

```
virtual int GetWidth() const { return m_Width; }
virtual int GetHeight() const { return m_Height; }
virtual int GetX() const { return m_X; }
virtual int GetY() const { return m_Y; }
```

Then, getters and setters for layout parameters follow:

```
virtual void SetAlignmentV( eAlignV V ) { m_AlignV = V; }
virtual void SetAlignmentH( eAlignH H ) { m_AlignH = H; }
virtual eAlignV GetAlignmentV() const { return m_AlignV; }
virtual eAlignH GetAlignmentH() const { return m_AlignH; }
virtual void SetParentFractionX( float X )
{ m_ParentFractionX = X; }
virtual void SetParentFractionY( float Y )
{ m_ParentFractionY = Y; }
```

The Add() and Remove() methods provide access to the m_ChildViews container:

```
virtual void Add( const clPtr<clUIView>& V )
{
  m_ChildViews.push_back( V );
}
virtual void Remove( const clPtr<clUIView>& V )
{
  m_ChildViews.erase(
    std::remove( m_ChildViews.begin(),
      m_ChildViews.end(), V ),
    m_ChildViews.end() );
}
```

Direct read-only access to m_ChildViews is provided by the GetChildViews() method:

```
virtual const std::vector< clPtr<clUIView> >&
  GetChildViews() const { return m_ChildViews; }
```

The Draw() method calls PreDrawView() to render the background layer of this UI element, then it calls Draw() for each and every child view, and finally, the call to PostDrawView() finishes the rendering process for this UI element:

```
virtual void Draw( const clPtr<iCanvas>& C )
{
  this->PreDrawView( C );
  for ( auto& i : m_ChildViews )
  {
    i->Draw( C );
```

```
    }
    this->PostDrawView( C );
}
```

The `UpdateScreenPositions()` method recalculates absolute screen positions of child views:

```
virtual void UpdateScreenPositions(
    int ParentX = 0, int ParentY = 0 )
{
    m_ScreenX = ParentX + m_X;
    m_ScreenY = ParentY + m_Y;
    for ( auto& i : m_ChildViews )
    {
        i->UpdateScreenPositions( m_ScreenX, m_ScreenY );
    }
}
```

The event handling part consists of `Update()` and `OnTouch()` methods. The `Update()` method informs all the child views a period of time has passed:

```
virtual void Update( double Delta )
{
    for( auto& i: m_ChildViews )
    i->Update( Delta );
}
```

The `OnTouch()` method accepts screen coordinates and a touch flag:

```
virtual bool OnTouch( int x, int y, bool Pressed )
{
    if ( IsPointOver( x, y ) )
    {
```

Check whether the touch event was handled by any of child views:

```
        for( auto& i: m_ChildViews )
        {
            if( i->OnTouch( x, y, Pressed ) )
            return true;
        }
    }
    return false;
}
```

The `IsPointOver()` method checks whether the point is inside the view:

```
virtual bool IsPointOver( int x, int y ) const
{
  return ( x >= m_ScreenX ) &&
    ( x <= m_ScreenX + m_Width  ) &&
    ( y >= m_ScreenY ) &&
    ( y <= m_ScreenY + m_Height );
}
```

The protected part contains two virtual methods to render the contents of the actual `clUIView`. The `PreDrawView()` method is called before rendering child views, so the visible result of this call may be erased by children, for example, a background layer. The `PostDrawView()` method is called after all child views had been rendered, like a decoration on top of the rendered image:

```
protected:
  virtual void PreDrawView( const clPtr<iCanvas>& C ) {};
  virtual void PostDrawView( const clPtr<iCanvas>& C ) {};
};
```

This mechanism enables UI rendering and customization. The last thing we need before our UI can come to life is an event dispatching mechanism. Let's implement it.

Events

At the lowest level, all the events from Android or desktop OSes are handled by the SDL library, and we only have to write the handler for these events:

```
bool clSDLWindow::HandleEvent( const SDL_Event& Event );
```

We add two more case labels to `HandleEvent()` so we can dispatch touch events:

```
case SDL_MOUSEBUTTONDOWN:
  OnTouch( Event.button.x, Event.button.y, true );
  break;
case SDL_MOUSEBUTTONUP:
  OnTouch( Event.button.x, Event.button.y, false );
  break;
```

Prior to C++11, wrapping C-like function pointers and class member function pointers in a single object was not an easy task requiring some heavy template library such as `boost::bind`. Now, the `std::function` object from the SDL library fits just fine for our purposes.

The only interactive object we implement here is `clUIButton`. When a user taps such an object, a custom action is performed. The code for the action can reside in a standalone function, a member function, or in a lambda expression. For example, we create an `Exit` button, and the code might be as follows:

```
ExitBtn->SetTouchHandler(
  [](int x, int y )
  {
    LOGI( "Exiting" );
    g_Window->RequestExit();
    return true;
  }
);
```

The `clUIButton` class must contain the `std::function` field, and the `OnTouch()` method optionally invokes this function when a tap occurs.

Implementing UI classes

The `clUIStatic` view is a descendant of `clUIView` with the overridden `PreDrawView()` method:

```
class clUIStatic: public clUIView
{
public:
  clUIStatic() : m_BackgroundColor( 255, 255, 255, 255 ) {}
  virtual void SetBackgroundColor( const ivec4& C )
  { m_BackgroundColor = C;};
protected:
  virtual void PreDrawView( const clPtr<iCanvas>& C ) override
  {
    C->SetColor( m_BackgroundColor );
    C->Rect(m_ScreenX, m_ScreenY, m_Width, m_Height, true);
    clUIView::PreDrawView( C );
  }
private:
  ivec4 m_BackgroundColor;
};
```

The `clUIButton` class adds a custom touch event handling atop of `clUIStatic` rendering:

```cpp
typedef std::function<bool(int x, int y)> sTouchHandler;
  class clUIButton: public clUIStatic
{
public:
  clUIButton(): m_OnTouchHandler(nullptr) {}
  virtual bool OnTouch( int x, int y, bool Pressed ) override
  {
    if( IsPointOver( x, y ) )
    {
      if(!Pressed && m_OnTouchHandler )
      return m_OnTouchHandler(x, y);
    }
    return false;
  }
  virtual void SetTouchHandler(const sTouchHandler&& H)
  { m_OnTouchHandler = H; }
private:
  sTouchHandler m_OnTouchHandler;
};
```

Now, our mini user interface can be used in applications.

Using views in application

Here is a short code snippet that creates a button and exits the application once this button is tapped:

```cpp
auto MsgBox = make_intrusive<clUIButton>();
MsgBox->SetParentFractionX( 0.5f );
MsgBox->SetParentFractionY( 0.5f );
MsgBox->SetAlignmentV( eAlignV_Center );
MsgBox->SetAlignmentH( eAlignH_Center );
MsgBox->SetBackgroundColor( ivec4( 255, 255, 255, 255) );
MsgBox->SetTitle("Exit");
MsgBox->SetTouchHandler( [](int x, int y )
  {
    LOGI( "Exiting" );
    g_Window->RequestExit();
    return true;
  }
);
```

The full source code is in the 1_SDL2UI example. Besides the details discussed in this chapter, the source code contains a basic layouting mechanism so that views can have relative coordinates and sizes. To get this bonus, take a look at LayoutController.cpp and LayoutController.h.

Summary

In this chapter, we learned how to implement and render basic user interface in C++, render UTF-8 text using the FreeType library, and handle user input in a platform independent way. We will use this functionality in the last chapter to implement a multiplatform gaming application. Now, let's return to the topic of 3D rendering started in *Chapter 6, OpenGL ES 3.1 and Cross-platform Rendering*, and implement a rendering engine on top of those abstractions.

8
Writing a Rendering Engine

In one of the previous chapters, we learned how to organize a thin abstraction layer on top of mobile and desktop OpenGL. Now, we can move into actual rendering territory and use this layer to implement a 3D rendering framework capable of rendering geometry loaded from files using materials, lights, and shadows.

The scene graph

A scene graph is a data structure commonly used to construct hierarchical representations of spatial graphical scenes. The main limitation of the classes introduced in *Chapter 6, OpenGL ES 3.1 and Cross-platform Rendering* is that they lack information on the scene as a whole. Users of these classes have to do ad hoc bookkeeping of transformations, state changes and dependencies, making implementation and support of any somewhat complex scenes a very challenging task. Furthermore, a lot of rendering optimizations cannot be done unless the whole scene information for the current frame is accessible.

In our current low-level implementation, we describe all visible entities using the clVertexArray class and render them using a shader program accessible via the clGLSLShaderProgram class with an ugly manual binding of matrices and shader parameters. Let's learn how to put all these properties together into a higher level data structure. First, we will start with a scene graph node.

The `clSceneNode` class contains local and global transformations and a vector of child nodes. These fields are protected, and we access them using setters and getters:

```
class clSceneNode: public iIntrusiveCounter
{
protected:
  mat4 m_LocalTransform;
  mat4 m_GlobalTransform;
  std::vector< clPtr<clSceneNode> > m_ChildNodes;
```

When we need hierarchy, we have to distinguish global and local transformation of the node. The local transformation is easy to understand from a user's point of view. This defines the position and orientation of a node relative to its parent node in a hierarchical spatial structure. The global transformation is used to render the geometry. *Per se*, it transforms the geometry from the model coordinate system into the world coordinate system. Local transformations can be intuitively modified by hand and global transformations should be subsequently reevaluated. The constructor of `clSceneNode` sets default transformation values:

```
public:
  clSceneNode():
  m_LocalTransform( mat4::Identity() ),
  m_GlobalTransform( mat4::Identity() ) {}
```

The `clSceneNode` class contains setters and getters to access and modify transformation matrices. The implementation is simple. However, it is quite cumbersome, so only the methods for the local transformation matrix are cited here. Check out the source code `1_SceneGraphRenderer` for the complete picture:

```
void SetLocalTransform( const mat4& Mtx )
{ m_LocalTransform = Mtx; }
const mat4& GetLocalTransformConst() const
{ return m_LocalTransform; }
mat4& GetLocalTransform()
{ return m_LocalTransform; };
```

Besides this, we need a method to add a child node to this scene node. Our current implementation is very simple:

```
virtual void Add( const clPtr<clSceneNode>& Node )
{
  m_ChildNodes.push_back( Node );
}
```

However, this method can be extended in the future to allow certain optimizations. For example, we can mark certain parts of a scene graph as dirty once we add new nodes. This will allow us to preserve inter-frame rendering queues constructed from the scene graph.

Sometimes, it is required that you set the global transformation matrix directly. For example, if you want to use a physics simulation library in your 3D application. Once done, local transformations should be recomputed. The `RecalculateLocalFromGlobal()` method calculates relative local transformations for each node in the hierarchy. For the root node, local and global transformations coincide. For the children, each global transformation matrix must be multiplied by the inverse global transformation of its parent:

```
void RecalculateLocalFromGlobal()
{
   mat4 ParentInv = m_GlobalTransform.GetInversed();
   for ( auto& i : m_ChildNodes )
   {
```

We multiply the parent's node global inverse transformation by the global transformation of the current node:

```
i->SetLocalTransform(
   ParentInv * i->GetGlobalTransform() );
```

The process is repeated down the hierarchy:

```
      i->RecalculateLocalFromGlobal();
   }
}
```

There is one more interesting thing left in the declaration of `clSceneNode`. This is the pure virtual method `AcceptTraverser()`. To render a scene graph, a technique known as the *visitor design pattern* is used (`https://en.wikipedia.org/?title=Visitor_pattern`):

```
virtual void AcceptTraverser(iSceneTraverser* Traverser) = 0;
```

The `iSceneTraverser` interface is used to define different operations on a scene graph. Since the number of different types of scene graph nodes is limited and constant, we can add new operations simply by implementing the `iSceneTraverser` interface:

```
class iSceneTraverser: public iIntrusiveCounter
{
public:
   virtual void Traverse( clPtr<clSceneNode> Node );
   virtual void Reset() = 0;
```

The interface is declared as a friend of all descendants of clSceneNode to allow direct access to the fields of these classes:

```
    friend class clSceneNode;
    friend class clMaterialNode;
    friend class clGeometryNode;
  protected:
    virtual void PreAcceptSceneNode( clSceneNode* Node ) {};
    virtual void PostAcceptSceneNode( clSceneNode* Node ) {};
    virtual void PreAcceptMaterialNode( clMaterialNode* Node ) {};
    virtual void PostAcceptMaterialNode( clMaterialNode* Node ){};
    virtual void PreAcceptGeometryNode( clGeometryNode* Node ) {};
    virtual void PostAcceptGeometryNode( clGeometryNode* Node ){};
  };
```

The implementation of Traverse() is shared between all traversers. It resets the traverser and calls the virtual method clSceneNode::AcceptTraverser():

```
  void iSceneTraverser::Traverse( clPtr<clSceneNode> Node )
  {
    if ( !Node ) return;
    Reset();
    Node->AcceptTraverser( this );
  }
```

In the declaration of iSceneTraverser, you can see two additional types of scene nodes. A tree of clSceneNode objects can hold a hierarchy of spatial transformations. Obviously, this is not enough to render anything yet. To do this, we need at least geometry data and a material.

Let's declare two more classes for this purpose: clMaterialNode and clGeometryNode.

For the first example of this chapter, a material will contain only ambient and diffuse colors (https://en.wikipedia.org/wiki/Phong_shading):

```
  struct sMaterial
  {
  public:
    sMaterial()
    : m_Ambient( 0.2f )
    , m_Diffuse( 0.8f )
    , m_MaterialClass()
    {}
    vec4 m_Ambient;
    vec4 m_Diffuse;
```

The field `m_MaterialClass` contains a material identifier, which can be used to distinguish special materials, for example, enable alpha transparency, for particle rendering:

```
    std::string m_MaterialClass;
};
```

Now, a material scene node can be declared. It is a simple data container:

```
class clMaterialNode: public clSceneNode
{
public:
  clMaterialNode() {};
  sMaterial& GetMaterial() { return m_Material; }
  const sMaterial& GetMaterial() const { return m_Material; }
  void SetMaterial( const sMaterial& Mtl ) { m_Material = Mtl; }
  virtual void AcceptTraverser(iSceneTraverser* Traverser) override;
private:
  sMaterial m_Material;
};
```

Let's take a look at the `AcceptTraverser()` method implementation. It is very simple and pretty efficient:

```
void clMaterialNode::AcceptTraverser( iSceneTraverser* Traverser )
{
  Traverser->PreAcceptSceneNode( this );
  Traverser->PreAcceptMaterialNode( this );
  AcceptChildren( Traverser );
  Traverser->PostAcceptMaterialNode( this );
  Traverser->PostAcceptSceneNode( this );
}
```

Geometry nodes are a bit more complex. This is because the API-independent geometry data representation in `clVertexAttribs` should be fed into the instance of `clGLVertexArray`:

```
class clGeometryNode: public clSceneNode
{
public:
  clGeometryNode() {};
  clPtr<clVertexAttribs> GetVertexAttribs() const
  { return m_VertexAttribs; }
  void SetVertexAttribs( const clPtr<clVertexAttribs>& VA )
  { m_VertexAttribs = VA; }
```

Here, we feed the geometry data into OpenGL in a lazy way:

```
clPtr<clGLVertexArray> GetVA() const
{
  if ( !m_VA )
  {
    m_VA = make_intrusive<clGLVertexArray>();
    m_VA->SetVertexAttribs( m_VertexAttribs );
  }
  return m_VA;
}
virtual void AcceptTraverser(iSceneTraverser* Traverser) override;
private:
  clPtr<clVertexAttribs> m_VertexAttribs;
  mutable clPtr<clGLVertexArray> m_VA;
};
```

The implementation of `AcceptTraverser()` is very similar to the one inside `clMaterialNode`. Just take a look into the bundled source code.

As you can see, the whole bunch of scene node classes is nothing but a simple data container. Actual operations happen in the traverser classes. The first implementation of `iSceneTraverser` is the `clTransformUpdateTraverser` class, which updates the global—which means relative to the root of the graph—transformation of each scene node:

```
class clTransformUpdateTraverser: public iSceneTraverser
{
private:
  std::vector<mat4> m_ModelView;
```

The private field `m_ModelView` contains a stack of matrices implemented as `std::vector`. The top element of this stack is the current global transformation of the node. The `Reset()` method clears the stack and pushes the identity matrix on the stack, which is later used as the global transformation of the root scene node:

```
public:
  clTransformUpdateTraverser(): m_ModelView() {}
  virtual void Reset() override
  {
    m_ModelView.clear();
    m_ModelView.push_back( mat4::Identity() );
  }
```

The `PreAcceptSceneNode()` method pushes a new value of the current global transformation onto the `m_ModelView` stack, and then uses it as the global transformation of incoming nodes:

```
protected:
  virtual void PreAcceptSceneNode( clSceneNode* Node ) override
  {
    m_ModelView.push_back( Node->GetLocalTransform() *
      m_ModelView.back() );
    Node->SetGlobalTransform( m_ModelView.back() );
  }
```

The `PostAcceptSceneNode()` method pops the topmost, now unused, matrix from the stack:

```
  virtual void PostAcceptSceneNode( clSceneNode* Node ) override
  {
    m_ModelView.pop_back();
  }
};
```

This traverser is used in the beginning of every frame before any geometry is rendered:

```
clTransformUpdateTraverser g_TransformUpdateTraverser;
clPtr<clSceneNode> g_Scene;
g_TransformUpdateTraverser.Traverse( g_Scene );
```

We are now almost ready to proceed with the actual rendering. To do this, we need to linearize the scene graph into a vector of rendering operations. Let's take a look into the `ROP.h` file. Each rendering operation is a freestanding piece of geometry, a material, and a bunch of transformation matrices. The required matrices are stored within the `sMatrices` structure:

```
struct sMatrices
{
```

The projection, view, and model matrices are set explicitly from the external state:

```
mat4 m_ProjectionMatrix;
mat4 m_ViewMatrix;
mat4 m_ModelMatrix;
```

Other matrices that are necessary for lighting and shading are updated using the `UpdateMatrices()` method:

```
mat4 m_ModelViewMatrix;
mat4 m_ModelViewMatrixInverse;
mat4 m_ModelViewProjectionMatrix;
mat4 m_NormalMatrix;
void UpdateMatrices()
{
  m_ModelViewMatrix = m_ModelMatrix * m_ViewMatrix;
  m_ModelViewMatrixInverse = m_ModelViewMatrix.GetInversed();
  m_ModelViewProjectionMatrix = m_ModelViewMatrix *
    m_ProjectionMatrix;
  m_NormalMatrix = mat4(
    m_ModelViewMatrixInverse.ToMatrix3().GetTransposed() );
}
};
```

This structure can be easily extended with additional matrices on an as-needed basis. Furthermore, it is very convenient to pack the values of this structure into a Uniform Buffer Object.

Now, our rendering operation can look as follows:

```
class clRenderOp
{
public:
  clRenderOp( const clPtr<clGeometryNode>& G,
    const clPtr<clMaterialNode>& M):
    m_Geometry(G), m_Material(M) {}
  void Render( sMatrices& Matrices ) const;
  clPtr<clGeometryNode> m_Geometry;
  clPtr<clMaterialNode> m_Material;
};
```

A minimalistic implementation of `clRenderOp::Render()` can be found in `ROP.cpp`. There are two shaders defined there. First, a generic vertex shader to transform normals into the world space:

```
static const char g_vShaderStr[] =
R"(
  uniform mat4 in_ModelViewProjectionMatrix;
  uniform mat4 in_NormalMatrix;
  uniform mat4 in_ModelMatrix;
```

```
  in vec4 in_Vertex;
  in vec2 in_TexCoord;
  in vec3 in_Normal;
  out vec2 v_Coords;
  out vec3 v_Normal;
  out vec3 v_WorldNormal;
  void main()
  {
    v_Coords = in_TexCoord.xy;
    v_Normal = mat3(in_NormalMatrix) * in_Normal;
    v_WorldNormal = ( in_ModelMatrix * vec4(
      in_Normal, 0.0 ) ).xyz;
    gl_Position = in_ModelViewProjectionMatrix * in_Vertex;
  }
)";
```

Then, a fragment shader that does simple per-pixel lighting using a single directional light source pointing in the same direction as the camera:

```
static const char g_fShaderStr[] =
R"(
  in vec2 v_Coords;
  in vec3 v_Normal;
  in vec3 v_WorldNormal;
  out vec4 out_FragColor;
  uniform vec4 u_AmbientColor;
  uniform vec4 u_DiffuseColor;
  void main()
  {
    vec4 Ka = u_AmbientColor;
    vec4 Kd = u_DiffuseColor;
```

The camera is statically positioned and lighting is done in the world space:

```
    vec3 L = normalize( vec3( 0.0, 0.0, 1.0 ) );
    float d = clamp( dot( L, normalize(v_WorldNormal) ),
      0.0, 1.0 );
    vec4 Color = Ka + Kd * d;
    out_FragColor = Color;
  }
)";
```

A static global variable holds a shader program linked using the shaders mentioned in the preceding code:

```
clPtr<clGLSLShaderProgram> g_ShaderProgram;
```

The actual rendering code updates all the matrices, sets parameters of the shader program and renders the geometry:

```
void clRenderOp::Render( sMatrices& Matrices ) const
{
  if ( !g_ShaderProgram )
  {
    g_ShaderProgram = make_intrusive<clGLSLShaderProgram>(
      g_vShaderStr, g_fShaderStr );
  }
  Matrices.m_ModelMatrix =
    this->m_Geometry->GetGlobalTransformConst();
  Matrices.UpdateMatrices();
```

The following piece of code will become a bottleneck once the number of rendering operations and uniforms increases. It can be improved using precached uniform locations:

```
  g_ShaderProgram->Bind();
  g_ShaderProgram->SetUniformNameVec4Array( "u_AmbientColor", 1,
    this->m_Material->GetMaterial().m_Ambient );
  g_ShaderProgram->SetUniformNameVec4Array( "u_DiffuseColor", 1,
    this->m_Material->GetMaterial().m_Diffuse );
  g_ShaderProgram->SetUniformNameMat4Array(
    in_ProjectionMatrix", 1, Matrices.m_ProjectionMatrix );
  g_ShaderProgram->SetUniformNameMat4Array( "in_ViewMatrix", 1,
    Matrices.m_ViewMatrix );
  g_ShaderProgram->SetUniformNameMat4Array( "in_ModelMatrix", 1,
    Matrices.m_ModelMatrix );
  g_ShaderProgram->SetUniformNameMat4Array( "in_NormalMatrix",
    1, Matrices.m_NormalMatrix );
  g_ShaderProgram->SetUniformNameMat4Array(
    "in_ModelViewMatrix", 1, Matrices.m_ModelViewMatrix );
  g_ShaderProgram->SetUniformNameMat4Array(
    "in_ModelViewProjectionMatrix", 1,
    Matrices.m_ModelViewProjectionMatrix );
  this->m_Geometry->GetVA()->Draw( false );
}
```

Let's convert a scene graph into a vector of rendering operations, so we can see the actual rendered image. This is done by the `clROPsTraverser` class:

```
class clROPsTraverser: public iSceneTraverser
{
private:
  std::vector<clRenderOp> m_RenderQueue;
  std::vector<clMaterialNode*> m_Materials;
public:
  clROPsTraverser()
  : m_RenderQueue()
  , m_Materials() {}
```

Clear everything before constructing the new queue of rendering operations:

```
virtual void Reset() override
{
  m_RenderQueue.clear();
  m_Materials.clear();
}
```

Return a reference to the most recently constructed queue:

```
virtual const std::vector<clRenderOp>& GetRenderQueue() const
{
  return m_RenderQueue;
}
```

Now, we implement the `iSceneTraverser` interface. Most of the methods here will be empty:

```
protected:
  virtual void PreAcceptSceneNode( clSceneNode* Node ) override
  {
  }
  virtual void PostAcceptSceneNode( clSceneNode* Node ) override
  {
  }
```

As the next geometry node comes in, use the topmost material from the materials stack and create a new rendering operation:

```
virtual void PreAcceptGeometryNode(
  clGeometryNode* Node ) override
{
```

```
      if ( !m_Materials.size() ) return;
      m_RenderQueue.emplace_back( Node, m_Materials.front() );
    }
    virtual void PostAcceptGeometryNode(
      clGeometryNode* Node ) override
    {
    }
```

The materials stack is updated on every incoming `clMaterialNode`:

```
    virtual void PreAcceptMaterialNode(
      clMaterialNode* Node ) override
    {
      m_Materials.push_back( Node );
    }
    virtual void PostAcceptMaterialNode(
      clMaterialNode* Node ) override
    {
      m_Materials.pop_back();
    }
  };
```

Finally, this framework can now be used to render the actual 3D graphics. The example scene is created in `1_SceneGraphRenderer/main.cpp`. First, the root of our scene is created:

```
  g_Scene = make_intrusive<clSceneNode>();
```

A red material is created and binded to a material scene node:

```
  auto MaterialNode = make_intrusive<clMaterialNode>();
  sMaterial Material;
  Material.m_Ambient = vec4( 0.1f, 0.0f, 0.0f, 1.0f );
  Material.m_Diffuse = vec4( 0.9f, 0.0f, 0.0f, 1.0f );
  MaterialNode->SetMaterial( Material );
```

Let's create a cube centered in the origin:

```
  auto VA =
    clGeomServ::CreateAxisAlignedBox( vec3(-1), vec3(+1) );
  g_Box= make_intrusive<clGeometryNode>();
  g_Box->SetVertexAttribs( VA );
  MaterialNode->Add( g_Box );
```

And add it in to the scene:

```
g_Scene->Add( MaterialNode );
```

Rendering is straightforward and very generic:

```
void OnDrawFrame()
{
   static float Angle = 0.0f;
```

Rotate the cube around its diagonal:

```
g_Box->SetLocalTransform(
   mat4::GetRotateMatrixAxis( Angle, vec3( 1, 1, 1 ) ) );
Angle += 0.01f;
```

Update global transformations of the nodes and construct a rendering queue:

```
g_TransformUpdateTraverser.Traverse( g_Scene );
g_ROPsTraverser.Traverse( g_Scene );
const auto& RenderQueue = g_ROPsTraverser.GetRenderQueue();
```

Set up the matrices. The camera has only a dummy implementation, which currently returns an identity view matrix:

```
sMatrices Matrices;
Matrices.m_ProjectionMatrix = Math::Perspective(
   45.0f, g_Window->GetAspect(), 0.4f, 2000.0f );
Matrices.m_ViewMatrix = g_Camera.GetViewMatrix();
```

Clear the screen before the frame will be rendered:

```
LGL3->glClear( GL_COLOR_BUFFER_BIT | GL_DEPTH_BUFFER_BIT );
LGL3->glEnable( GL_DEPTH_TEST );
```

Iterate over the rendering queue and render everything:

```
for ( const auto& ROP : RenderQueue )
{
   ROP.Render( Matrices );
}
}
```

The resulting image will look like the following screenshot:

Let's now extend our scene graph rendering example with lights and shadows, and make sure everything works on Android.

Lighting and shading

To render lights and shadows, we need to extend the approach shown in the previous paragraphs. The next code example we will discuss is `2_ShadowMaps`. Shadow mapping will be done using projected shadow maps (https://en.wikipedia.org/wiki/Shadow_mapping). This way a scene is rendered into an off-screen depth buffer from the light's point of view. Next, the scene is rendered as usual and every fragment is projected onto the lights shadow map and the depth value relative to the light is compared to the value in the constructed shadow map. The fragment is in shadow if the depth value is larger than the corresponding depth value from the shadow map. To do an off-screen rendering, we need to revisit our OpenGL wrapper introduced in *Chapter 6, OpenGL ES 3.1 and Cross-platform Rendering*, and add a framebuffer abstraction to it.

The `clGLFrameBuffer` class is declared in `GLFrameBuffer.h`:

```
class clGLFrameBuffer: public iIntrusiveCounter
{
public:
  clGLFrameBuffer():
  FFrameBuffer(0),
  FColorBuffer(),
  FDepthBuffer(),
  FColorBuffersParams(),
  FHasDepthBuffer( false )
  {}
  virtual ~clGLFrameBuffer();
```

The method `InitRenderTargetV()` accepts a vector containing integer values of width, height, and bits per channel:

```
virtual void InitRenderTargetV(
    const ivec4& WidthHeightBitsPerChannel,
    const bool HasDepthBuffer );
```

This method provides access to the private data members, those are width, height, and bits per channel, which were passed into `InitRenderTargetV()`:

```
virtual ivec4 GetParameters() const
{
   return FColorBuffersParams;
}
```

The most important ability of a framebuffer is being able to provide its content as textures—a color texture and a depth texture:

```
virtual clPtr<clGLTexture> GetColorTexture() const
{
   return FColorBuffer;
}
virtual clPtr<clGLTexture> GetDepthTexture() const
{
   return FDepthBuffer;
}
```

The `Bind()` method sets this framebuffer as the current OpenGL framebuffer:

```
virtual void Bind( int TargetIndex ) const;
virtual void UnBind() const;
```

The protected method `CheckFrameBuffer()` is used to check the completeness of the frame buffer according to the OpenGL specification:

```
protected:
   void CheckFrameBuffer() const;
```

The private section of the class contains an OpenGL buffer identifier, two `clGLTexture` objects for color and depth textures, respectively, and two fields containing framebuffer parameters:

```
private:
   GLuint FFrameBuffer;
   clPtr<clGLTexture> FColorBuffer;
   clPtr<clGLTexture> FDepthBuffer;
   ivec4 FColorBuffersParams;
   bool FHasDepthBuffer;
};
```

Correct construction of a framebuffer for Android and other platforms requires some work and careful selection of parameters. Let's take a look at the implementation of the `InitRenderTargetV()` member function:

```
void clGLFrameBuffer::InitRenderTargetV(
   const ivec4& WidthHeightBitsPerChannel,
   const bool HasDepthBuffer )
{
```

At first, we store framebuffer's parameters in private data members. These values are used later for viewport aspect computation:

```
FColorBuffersParams = WidthHeightBitsPerChannel;
FHasDepthBuffer = HasDepthBuffer;
```

Next, we will call the OpenGL function to create a framebuffer object:

```
LGL3->glGenFramebuffers( 1, &FFrameBuffer );
```

Once the framebuffer object is created, we can bind it as the current framebuffer to set up its properties:

```
Bind( 0 );
```

Create and attach a color texture to the framebuffer:

```
FColorBuffer = make_intrusive<clGLTexture>();
int Width = FColorBuffersParams[0];
int Height = FColorBuffersParams[1];
```

The only difference here between Android and desktop implementations is the selection of buffer data formats. OpenGL 4 Core Profile requires the number of bits in the internal format and depth format to be specified explicitly while OpenGL ES 3 wants generic `GL_RGBA` and `GL_DEPTH_COMPONENT` respectively. We declare two constants in a platform-specific way:

```
#if defined( OS_ANDROID )
   const Lenum InternalFormat = GL_RGBA;
   auto DepthFormat = GL_DEPTH_COMPONENT;
#else
   const Lenum InternalFormat = GL_RGBA8;
   auto DepthFormat = GL_DEPTH_COMPONENT24;
#endif
```

We will call the `SetFormat()` method of `clGLTexture` to set up the format of the color texture:

```
FColorBuffer->SetFormat(
   GL_TEXTURE_2D, InternalFormat, GL_RGBA, Width, Height );
```

The `AttachToCurrentFB()` methods attaches the created color texture to the currently binded framebuffer. The value of `GL_COLOR_ATTACHMENT0` specifies an OpenGL attachment point:

```
FColorBuffer->AttachToCurrentFB( GL_COLOR_ATTACHMENT0 );
```

A shadow map contains depth buffer values, so we create a depth texture the following way on an as-needed basis:

```
if ( HasDepthBuffer )
{
   FDepthBuffer = make_intrusive<clGLTexture>();
```

The depth buffer should have the same dimensions as the color buffer:

```
int Width = FColorBuffersParams[0];
int Height = FColorBuffersParams[1];
```

The setup of the depth buffer is similar to that of the color buffer:

```
FDepthBuffer->SetFormat( GL_TEXTURE_2D,
   DepthFormat, GL_DEPTH_COMPONENT, Width, Height );
FDepthBuffer->AttachToCurrentFB( GL_DEPTH_ATTACHMENT );
}
```

To ensure correct operation, we will check the error code and unbind the buffer:

```
CheckFrameBuffer();
UnBind();
}
```

The `CheckFrameBuffer()` member function uses OpenGL calls to check the current state of a framebuffer:

```
void clGLFrameBuffer::CheckFrameBuffer() const
{
Bind( 0 );
GLenum FBStatus =
  LGL3->glCheckFramebufferStatus( GL_FRAMEBUFFER );
```

Convert an error code into a string and print it into the system log:

```
switch ( FBStatus )
{
  case GL_FRAMEBUFFER_COMPLETE: // Everything's OK
    break;
  case GL_FRAMEBUFFER_INCOMPLETE_ATTACHMENT:
    LOGI( "GL_FRAMEBUFFER_INCOMPLETE_ATTACHMENT" );
    break;
  case GL_FRAMEBUFFER_INCOMPLETE_MISSING_ATTACHMENT:
    LOGI("GL_FRAMEBUFFER_INCOMPLETE_MISSING_ATTACHMENT" );
    break;
```

OpenGL ES is missing some of the capabilities of OpenGL. Here, we omit them to make the code portable:

```
#if !defined(OS_ANDROID)
  case GL_FRAMEBUFFER_INCOMPLETE_DRAW_BUFFER:
    LOGI( "GL_FRAMEBUFFER_INCOMPLETE_DRAW_BUFFER" );
    break;
  case GL_FRAMEBUFFER_INCOMPLETE_READ_BUFFER:
    LOGI( "GL_FRAMEBUFFER_INCOMPLETE_READ_BUFFER" );
    break;
#endif
  case GL_FRAMEBUFFER_UNSUPPORTED:
    LOGI( "GL_FRAMEBUFFER_UNSUPPORTED" );
    break;
  default:
    LOGI( "Unknown framebuffer error: %x", FBStatus );
}
```

By default, nothing is printed:

```
    UnBind();
  }
```

Let's proceed further so that we can make use of this class.

Lights and light nodes

It is very convenient to represent a light source as a part of a 3D scene. When we write "a 3D scene", we mean a scene graph. In order to attach a light source to a scene graph, we need a special node for it. Here is the `clLightNode` class that holds a pointer to `iLight` with all light properties:

```
    class clLightNode: public clSceneNode
    {
    public:
      clLightNode() {}
      clPtr<iLight> GetLight() const
      {
        return m_Light;
      }
      void SetLight( const clPtr<iLight>& L )
      {
        m_Light = L;
      }
      virtual void AcceptTraverser( iSceneTraverser* Traverser )
        override;
      clPtr<iLight> m_Light;
    };
```

The `AcceptTraverser()` method is similar to those in `clGeometryNode` and `clMaterialNode`. But this time, we will call the `PreAcceptLightNode()` and `PostAcceptLightNode()` methods of `iSceneTraverser`:

```
    void clLightNode::AcceptTraverser( iSceneTraverser* Traverser )
    {
      Traverser->PreAcceptSceneNode( this );
      Traverser->PreAcceptLightNode( this );
      AcceptChildren( Traverser );
      Traverser->PostAcceptLightNode( this );
      Traverser->PostAcceptSceneNode( this );
    }
```

This new scene node type forces us to extend the interface of `iSceneTraverser`:

```
protected:
friend class clLightNode;
virtual void PreAcceptLightNode( clLightNode* Node ) {}
virtual void PostAcceptLightNode( clLightNode* Node ) {}
```

Traversers can now handle light nodes in a special way. We will use this ability to maintain a list of active lights within the scene on a per-frame basis.

The `iLight` class encapsulates light parameters. It calculates the required projection and view matrices of a light source, passes them into a shader program and holds a shadow map. We should note that holding an initialized shadow map for possible unused light source is certainly non-optimal. The least we can do in our minimalistic example is to postpone the creation of a shadow map to the moment when it is really needed. In our example, we will deal only with spotlights. However, this approach can be easily extended with directional and point lights:

```
class iLight: public iIntrusiveCounter
{
public:
   iLight() :
   m_Ambient(0.2f),
   m_Diffuse(0.8f),
   m_Position(0),
   m_Direction(0.0f, 0.0f, 1.0f),
```

In case you want to implement multiple types of lights, it is advisable to push this field down to a class representing a spot light. Since our example has only lights of a single type, putting this value here is a reasonable compromise:

```
   m_SpotOuterAngle(45.0f)
   {}
```

The `UpdateROPUniforms()` method updates all the uniforms within a shader program required for shadow map rendering. The `clMaterialSystem` class is described in details after we finish with `iLight`:

```
void UpdateROPUniforms(
   const std::vector<clRenderOp>& ROPs,
   const clPtr<clMaterialSystem>& MatSys,
   const clPtr<clLightNode>& LightNode ) const;
```

To render the scene from the light's point of view, we need to calculate two matrices. The first one is the standard *look-at* matrix defining the light's view matrix and the second one is a perspective projection matrix defining the light's frustum:

```
mat4 GetViewForShadowMap() const
{
  return Math::LookAt(
    m_Position, m_Position + m_Direction,
    vec3( 0, 0, 1 ) );
}
mat4 GetProjectionForShadowMap() const
{
  float NearCP = 0.1f;
  float FarCP = 1000.0f;
  return Math::Perspective( 2.0f * this->m_SpotOuterAngle,
    1.0f, NearCP, FarCP );
}
```

The `GetShadowMap()` function returns a lazy-initialized shadow map buffer attached to this light source:

```
clPtr<clGLFrameBuffer> iLight::GetShadowMap() const
{
  if ( !m_ShadowMap )
  {
    m_ShadowMap = make_intrusive<clGLFrameBuffer>();
    m_ShadowMap->InitRenderTargetV(
      { ivec4(1024, 1024, 8, 0) },
      true
    );
  }
  return m_ShadowMap;
}
```

The properties of the light source include its diffuse and ambient colors used in a simple lighting model, position, and direction for viewing matrix calculation and the spot light cone angle:

```
public:
  vec4 m_Ambient;
  vec4 m_Diffuse;
  vec3 m_Position;
  vec3 m_Direction;
  float m_SpotOuterAngle;
```

In the end, we declare a framebuffer holding a shadow map:

```
mutable clPtr<clGLFrameBuffer> m_ShadowMap;
};
```

Let's see how uniforms of a shader program are updated. This happens in the `UpdateROPUniforms()`, which is called for every render operation before each shadow map is rendered:

```
void iLight::UpdateROPUniforms(
   const std::vector<clRenderOp>& ROPs,
   const clPtr<clMaterialSystem>& MatSys,
   const clPtr<clLightNode>& LightNode ) const
{
   mat4 LightMV = this->GetViewForShadowMap();
   mat4 LightProj = GetProjectionForShadowMap();
   mat4 Mtx = LightNode->GetGlobalTransformConst();
   vec3 Pos = ( Mtx * vec4( this->m_Position, 1.0f ) ).XYZ();
   vec3 Dir = ( Mtx * vec4( this->m_Direction, 0.0f ) ).XYZ();
   auto AmbientSP =
      MatSys->GetShaderProgramForPass( ePass_Ambient );
   AmbientSP->Bind();
   AmbientSP->SetUniformNameVec3Array( "u_LightPos", 1, Pos );
   AmbientSP->SetUniformNameVec3Array( "u_LightDir", 1, Dir );
   auto LightSP =
      MatSys->GetShaderProgramForPass( ePass_Light );
   LightSP->Bind();
   LightSP->SetUniformNameVec3Array( "u_LightPos", 1, Pos );
   LightSP->SetUniformNameVec3Array( "u_LightDir", 1, Dir );
   LightSP->SetUniformNameVec4Array(
      "u_LightDiffuse", 1, this->m_Diffuse );
   mat4 ScaleBias = GetProjScaleBiasMat();
   mat4 ShadowMatrix = ( Mtx * LightMV * LightProj ) * ScaleBias;
   LightSP->SetUniformNameMat4Array(
      "in_ShadowMatrix", 1, ShadowMatrix );
   this->GetShadowMap()->GetDepthTexture()->Bind( 0 );
}
```

The `GetProjScaleBiasMat()` helper routine returns a scaling matrix, which maps [-1..1] normalized device coordinates to the [0..1] range:

```
mat4 GetProjScaleBiasMat()
{
   return mat4(
      0.5f, 0.0f, 0.0f, 0.0f,
      0.0f, 0.5f, 0.0f, 0.0f,
      0.0f, 0.0f, 0.5f, 0.0f,
      0.5f, 0.5f, 0.5f, 1.0 );
}
```

The `clMaterialSystem` class mentioned in this code requires some additional explanation.

Material system

In our previous example, `1_SceneGraphRenderer`, we used a single shader program to render all objects in the scene. Now, our renderer will become multipass. We need to create shadow maps, and then render shadowed objects and calculate lighting. This is done using three different shader programs in three different rendering passes. To distinguish between passes, we define the `ePass` enum as follows:

```
enum ePass
{
  ePass_Ambient,
  ePass_Light,
  ePass_Shadow
};
```

To handle different shader programs based on passes and material properties, we implement the `clMaterialSystem` class:

```
class clMaterialSystem: public iIntrusiveCounter
{
public:
  clMaterialSystem();
```

The `GetShaderProgramForPass()` method returns the shader program for the specified pass stored in `std::map`:

```
clPtr<clGLSLShaderProgram> GetShaderProgramForPass(ePass P)
{
  return m_ShaderPrograms[ P ];
}
private:
  std::map<ePass, clPtr<clGLSLShaderProgram>> m_ShaderPrograms;
};
```

The constructor of this class creates each shader program required for rendering and inserts it into the map:

```
clMaterialSystem::clMaterialSystem()
{
  m_ShaderPrograms[ ePass_Ambient ] =
    make_intrusive<clGLSLShaderProgram>( g_vShaderStr,
      g_fShaderAmbientStr );
```

```
m_ShaderPrograms[ ePass_Light   ] =
  make_intrusive<clGLSLShaderProgram>( g_vShaderStr,
    g_fShaderLightStr );
m_ShaderPrograms[ ePass_Shadow  ] =
  make_intrusive<clGLSLShaderProgram>( g_vShaderShadowStr,
    g_fShaderShadowStr );
}
```

 In this example, a map can be replaced with a simple C-style array. However, later on, we will use different material types with different shader programs, so a map would fit just right.

As in the previous example, the source code of each shader is stored in a static string variable. This time, the code is a bit more complicated. The vertex shader source code is shared between the ambient and per-light passes:

```
static const char g_vShaderStr[] =
R"(
  uniform mat4 in_ModelViewProjectionMatrix;
  uniform mat4 in_NormalMatrix;
  uniform mat4 in_ModelMatrix;
  uniform mat4 in_ShadowMatrix;
  in vec4 in_Vertex;
  in vec2 in_TexCoord;
  in vec3 in_Normal;
  out vec2 v_Coords;
  out vec3 v_Normal;
  out vec3 v_WorldNormal;
  out vec4 v_ProjectedVertex;
  out vec4 v_ShadowMapCoord;
```

The same function was used in the C++ code to transform values from the [-1..1] to [0..1] range:

```
mat4 GetProjScaleBiasMat()
{
  return mat4(
    0.5, 0.0, 0.0, 0.0,
    0.0, 0.5, 0.0, 0.0,
    0.0, 0.0, 0.5, 0.0,
    0.5, 0.5, 0.5, 1.0 );
}
```

Values are passed to the subsequent fragment shaders:

```
void main()
{
  v_Coords = in_TexCoord.xy;
  v_Normal = mat3(in_NormalMatrix) * in_Normal;
  v_WorldNormal =
    ( in_ModelMatrix * vec4( in_Normal, 0.0 ) ).xyz;
  v_ProjectedVertex =
    GetProjScaleBiasMat() *
    in_ModelViewProjectionMatrix * in_Vertex;
  v_ShadowMapCoord =
    in_ShadowMatrix * in_ModelMatrix * in_Vertex;
  gl_Position = in_ModelViewProjectionMatrix * in_Vertex;
}
)";
```

Here is the fragment shader for the ambient pass. Just output the ambient color to the framebuffer, and we are done:

```
static const char g_fShaderAmbientStr[] =
R"(
  in vec2 v_Coords;
  in vec3 v_Normal;
  in vec3 v_WorldNormal;
  out vec4 out_FragColor;
  uniform vec4 u_AmbientColor;
  uniform vec4 u_DiffuseColor;
  void main()
  {
    out_FragColor = u_AmbientColor;
  }
)";
```

The fragment shader for per-light passes computes the actual lighting and shading based on light's parameters and the shadow map. This is why it is so long compared to all of our previous shaders:

```
static const char g_fShaderLightStr[] =
R"(
  in vec2 v_Coords;
  in vec3 v_Normal;
  in vec3 v_WorldNormal;
  in vec4 v_ProjectedVertex;
```

```
in vec4 v_ShadowMapCoord;
out vec4 out_FragColor;
uniform vec4 u_AmbientColor;
uniform vec4 u_DiffuseColor;
uniform vec3 u_LightPos;
uniform vec3 u_LightDir;
uniform vec4 u_LightDiffuse;
uniform sampler2D Texture0;
```

Shadows are computed using the technique called *percentage closer filtering*. If we use a naïve shadow mapping approach, the resulting shadows will have a lot of aliasing. The idea of **percentage closer filtering (PCF)** is to sample from the shadow map around the current pixel and compare its depth to all the samples. By averaging the results of comparison (not the results on the sampling), we can get smoother edges between light and shadow. Our example uses a 5 X 5 PCF filter with 26 taps:

```
float PCF5x5( const vec2 ShadowCoord, float Depth )
{
  float Size = 1.0 / float( textureSize( Texture0, 0 ).x );
  float Shadow =
    ( Depth >= texture( Texture0, ShadowCoord ).r ) ? 1.0 : 0.0;
  for ( int v=-2; v<=2; v++ ) for ( int u=-2; u<=2; u++ )
  {
    Shadow +=
      ( Depth >= texture( Texture0, ShadowCoord + Size *
        vec2(u, v) ).r ) ? 1.0 : 0.0;
  }
  return Shadow / 26.0;
}
```

Here's the function to evaluate whether a given fragment is in shadow:

```
float ComputeSpotLightShadow()
{
```

Do a perspective division to project the shadow map onto the object:

```
  vec4 ShadowCoords4 =
    v_ShadowMapCoord / v_ShadowMapCoord.w;
  if ( ShadowCoords4.w > 0.0 )
  {
    vec2 ShadowCoord = vec2( ShadowCoords4 );
    float DepthBias = -0.0002;
    float ShadowSample = 1.0 - PCF5x5( ShadowCoord,
      ShadowCoords4.z + DepthBias );
```

The `DepthBias` coefficient is used to prevent shadow acne. Here are two renderings of the same scene with the zero `DepthBias` (left) and `-0.0002` (right):

In general, it requires manual tweaking and should be a part of light's parameters. Take a look at the following link for more ideas on how to improve shadows:

`https://msdn.microsoft.com/en-us/library/windows/desktop/`
`ee416324(v=vs.85).aspx`.

Now, multiply the coefficients and return the resulting value:

```
      float ShadowCoef = 0.3;
      return ShadowSample * ShadowCoef;
   }
   return 1.0;
}
```

Now we can compute a simple lighting model based on the actual light direction and its shadow map:

```
   void main()
   {
      vec4 Kd = u_DiffuseColor * u_LightDiffuse;
```

Writing a Rendering Engine

```
    vec3 L = normalize( u_LightDir );
    vec3 N = normalize( v_WorldNormal );
    float d = clamp( dot( -L, N ), 0.0, 1.0 );
    vec4 Color = Kd * d * ComputeSpotLightShadow();
    Color.w = 1.0;
    out_FragColor = Color;
  }
)";
```

To construct a shadow map used in the previous shader, we need an additional rendering pass. For each light, the following vertex and fragment shaders are used:

```
static const char g_vShaderShadowStr[] =
R"(
  uniform mat4 in_ModelViewProjectionMatrix;
  in vec4 in_Vertex;
  void main()
  {
    gl_Position = in_ModelViewProjectionMatrix * in_Vertex;
  }
)";
static const char g_fShaderShadowStr[] =
R"(
  out vec4 out_FragColor;
  void main()
  {
    out_FragColor = vec4( 1, 1, 1, 1 );
  }
)";
```

Now we can render a nicer image with all shadows and much more accurate lighting. Let's take a look at the 2_ShadowMaps/main.cpp file.

Demo application and a rendering technique

The most important part of the new code is in the OnDrawFrame() method. It uses the clForwardRenderingTechnique class to render the scene. Let's take a look at Technique.cpp.

A helper function `RenderROPs()` is used to render a vector of rendering operations:

```
void RenderROPs(
  sMatrices& Matrices, const std::vector<clRenderOp>&
  RenderQueue, ePass Pass )
{
  for ( const auto& ROP : RenderQueue )
  {
    ROP.Render( Matrices, g_MatSys, Pass );
  }
}
```

Now, all the passes can be described in terms of this function. Take a look at the function `clForwardRenderingTechnique::Render()`. First, let's construct two rendering queues, for opaque and transparent objects. Transparent objects are those with the string `Particle` as their material class. We will make use of transparent objects in the next chapter:

```
m_TransformUpdateTraverser.Traverse( Root );
m_ROPsTraverser.Traverse( Root );
const auto& RenderQueue = m_ROPsTraverser.GetRenderQueue();
auto RenderQueue_Opaque =
  Select( RenderQueue, [] ( const clRenderOp& ROP )
    {
      return ROP.m_Material->GetMaterial().m_MaterialClass !=
        "Particle";
    } );
    auto RenderQueue_Transparent =
      Select( RenderQueue, [] ( const clRenderOp& ROP )
    {
      return ROP.m_Material->GetMaterial().m_MaterialClass ==
        "Particle";
    }
);
```

Prepare matrices for shaders and clear OpenGL buffers:

```
sMatrices Matrices;
Matrices.m_ProjectionMatrix = Proj;
Matrices.m_ViewMatrix = View;
Matrices.UpdateMatrices();
LGL3->glClearColor( 0.0f, 0.0f, 0.0f, 0.0f );
LGL3->glClear(
  GL_COLOR_BUFFER_BIT | GL_DEPTH_BUFFER_BIT );
LGL3->glEnable( GL_DEPTH_TEST );
```

Now, render all the objects with their ambient color. This is it, the ambient pass does not require any lights. As a by-product, we will have a Z-buffer filled with values, so we can disable depth writes in the subsequent passes:

```
LGL3->glDepthFunc( GL_LEQUAL );
LGL3->glDisablei( GL_BLEND, 0 );
RenderROPs( Matrices, RenderQueue_Opaque,
  ePass_Ambient, MatSys );
```

For the subsequent per-light passes, we need a vector of lights from the scene. Get it from the traverser and update all the shadow maps:

```
auto Lights = g_ROPsTraverser.GetLights();
UpdateShadowMaps( Lights, RenderQueue );
```

The `UpdateShadowMaps()` function iterates over a vector of light nodes and renders shadow casters into the corresponding shadow maps:

```
void UpdateShadowMaps(
  const std::vector<clLightNode*>& Lights,
  const std::vector<clRenderOp>& ROPs )
{
  LGL3->glDisable( GL_BLEND );
  for ( size_t i = 0; i != Lights.size(); i++ )
  {
    sMatrices ShadowMatrices;
    clPtr<iLight> L = Lights[ i ]->GetLight();
    clPtr<clGLFrameBuffer> ShadowBuffer = L->GetShadowMap();
```

Bind and clear the shadow map framebuffer:

```
    ShadowBuffer->Bind( 0 );
    LGL3->glClearColor( 0, 0, 0, 1 );
    LGL3->glClear(
      GL_COLOR_BUFFER_BIT | GL_DEPTH_BUFFER_BIT );
```

The light knows its projection and view matrices. This code is quite generic to be extended for use with light types, including lights with multiple viewing frustums:

```
    LMatrix4 Proj = L->GetProjectionForShadowMap();
    LMatrix4 MV = L->GetViewForShadowMap();
```

Update uniforms within the shader program:

```
ShadowMatrices.m_ViewMatrix = MV;
ShadowMatrices.m_ProjectionMatrix = Proj;
ShadowMatrices.UpdateMatrices();
L->UpdateROPUniforms( ROPs, g_MatSys, Lights[i] );
```

Render into the shadow map and unbind the framebuffer:

```
RenderROPs( ShadowMatrices, ROPs, ePass_Shadow );
ShadowBuffer->UnBind();
    }
  }
```

All the shadow maps are now ready to be used in the rendering code. Let's continue with the OnDrawFrame() function. Per-light passes accumulate lighting from all light sources and look as follows:

```
LGL3->glDepthFunc( GL_EQUAL );
LGL3->glBlendFunc( GL_ONE, GL_ONE );
LGL3->glEnablei( GL_BLEND, 0 );
for ( const auto& L : Lights )
{
    L->GetLight()->UpdateROPUniforms( RenderQueue, MatSys, L );
    RenderROPs( Matrices, RenderQueue, ePass_Light, MatSys );
}
```

Last but not the least, render the ambient lighting for transparent objects:

```
LGL3->glBlendFunc(GL_SRC_ALPHA, GL_ONE);
LGL3->glDepthFunc(GL_LESS);
LGL3->glEnablei(GL_BLEND, 0);
LGL3->glDepthMask( GL_FALSE );
RenderROPs( Matrices, RenderQueue_Transparent,
    ePass_Ambient, MatSys );
```

Don't forget to reset the OpenGL state. A good idea to extend the renderer would be to encapsulate states such as depth test, depth mask, blending mode, and others into a pipeline state object and update pipeline states only once they are changed. If you want to extend the examples into a full-scale rendering code, this improvement is a must-have:

```
LGL3->glDepthMask( GL_TRUE );
```

We covered all the low-level rendering code. Let's go one level higher and see how a scene can be constructed.

Scene construction

Our test scene is constructed in `main()` and the process looks the following way. First, global objects are instantiated:

```
g_MatSys = make_intrusive<clMaterialSystem>();
g_Scene = make_intrusive<clSceneNode>();
g_Canvas = make_intrusive<clGLCanvas>( g_Window );
```

After this, materials and material nodes are set up:

```
auto CubeMaterialNode = make_intrusive<clMaterialNode>();
{
  sMaterial Material;
  Material.m_Ambient = vec4( 0.2f, 0.0f, 0.0f, 1.0f );
  Material.m_Diffuse = vec4( 0.8f, 0.0f, 0.0f, 1.0f );
  CubeMaterialNode->SetMaterial( Material );
}
auto PlaneMaterialNode = make_intrusive<clMaterialNode>();
{
  sMaterial Material;
  Material.m_Ambient = vec4( 0.0f, 0.2f, 0.0f, 1.0f );
  Material.m_Diffuse = vec4( 0.0f, 0.8f, 0.0f, 1.0f );
  PlaneMaterialNode->SetMaterial( Material );
}
auto DeimosMaterialNode = make_intrusive<clMaterialNode>();
{
  sMaterial Material;
  Material.m_Ambient = vec4( 0.0f, 0.0f, 0.2f, 1.0f );
  Material.m_Diffuse = vec4( 0.0f, 0.0f, 0.8f, 1.0f );
  DeimosMaterialNode->SetMaterial( Material );
}
```

Now, we can create geometry of the scene using a bunch of boxes and a 3D model of Deimos (https://en.wikipedia.org/wiki/Deimos_(moon)) loaded from an .obj file:

```
{
  auto VA = clGeomServ::CreateAxisAlignedBox(
    vec3(-0.5), vec3(+0.5) );
  g_Box= make_intrusive<clGeometryNode>();
  g_Box->SetVertexAttribs( VA );
  CubeMaterialNode->Add( g_Box );
```

This function can be found in the `Loader_OBJ.cpp` file and does parsing of Wavefront OBJ file format (`https://en.wikipedia.org/wiki/Wavefront_.obj_file`):

```
    auto DeimosNode = LoadOBJSceneNode(
      g_FS->CreateReader( "deimos.obj" ) );
    DeimosNode->SetLocalTransform(
      mat4::GetScaleMatrix(vec3(0.01f, 0.01f, 0.01f)) *
      mat4::GetTranslateMatrix(vec3(1.1f, 1.1f, 0.0f))
    );
    DeimosMaterialNode->Add( DeimosNode );
  }
  {

    auto VA = clGeomServ::CreateAxisAlignedBox(
      vec3(-2.0f, -2.0f, -1.0f), vec3(2.0f, 2.0f, -0.95f) );
    auto Geometry = make_intrusive<clGeometryNode>();
    Geometry->SetVertexAttribs( VA );
    PlaneMaterialNode->Add( Geometry );
  }
```

And last but not least, we will add two lights to the scene, which will produce two distinct shadows:

```
  {
    auto Light = make_intrusive<iLight>( );
    Light->m_Diffuse = vec4( 0.5f, 0.5f, 0.5f, 1.0f );
    Light->m_Position = vec3( 5, 5, 5 );
    Light->m_Direction = vec3( -1, -1, -1 ).GetNormalized();
    Light->m_SpotOuterAngle = 21;
    g_LightNode = make_intrusive<clLightNode>( );
    g_LightNode->SetLight( Light );
    g_Scene->Add( g_LightNode );
  }
  {
    auto Light = make_intrusive<iLight>();
    Light->m_Diffuse = vec4( 0.5f, 0.5f, 0.5f, 1.0f );
    Light->m_Position = vec3( 5, -5, 5 );
    Light->m_Direction = vec3( -1, 1, -1 ).GetNormalized();
    Light->m_SpotOuterAngle = 20;
    auto LightNode = make_intrusive<clLightNode>();
    LightNode->SetLight( Light );
    g_Scene->Add( LightNode );
  }
```

Assemble it all together and proceed to the application main loop:

```
g_Scene->Add( CubeMaterialNode );
g_Scene->Add( PlaneMaterialNode );
g_Scene->Add( DeimosMaterialNode );
while( g_Window && g_Window->HandleInput() )
{
  OnDrawFrame();
  g_Window->Swap();
}
```

The resulting application renders the following image with a rotating cube and shadows from two light sources:

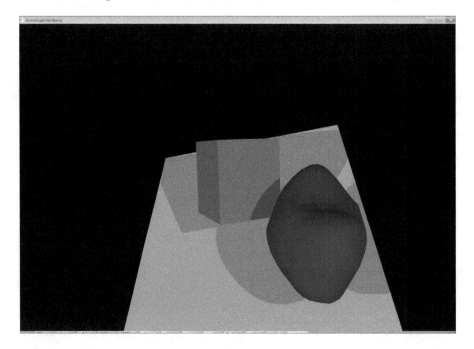

The demo application is runnable on Android as well. Just go and try it!

User interaction with 3D scenes

We hope that you tried to run the 2_ShadowMaps example. As you may have noticed, the 3D scene can be rotated with a gesture on a touch screen or using mouse on a desktop machine.

It is done using the `clVirtualTrackball` class, which emulates a virtual trackball by calculating a view matrix based on the provided touch points:

```
class clVirtualTrackball
{
public:
  clVirtualTrackball():
  FCurrentPoint( 0.0f ),
  FPrevPoint( 0.0f ),
  FStarted( false )
  {
    FRotation.IdentityMatrix();
    FRotationDelta.IdentityMatrix();
  };
```

Get the view matrix, corresponding to the new touch point:

```
virtual LMatrix4 DragTo(
  LVector2 ScreenPoint, float Speed, bool KeyPressed )
{
  if ( KeyPressed && !FStarted )
  {
    StartDragging( ScreenPoint );
    FStarted = KeyPressed;
    return mat4::Identity();
  }
  FStarted = KeyPressed;
```

If we are not touching the screen, return an identity matrix:

```
  if ( !KeyPressed ) return mat4::Identity();
```

Project the touch point onto the virtual trackball sphere and find the distance between the current and the previous projection points:

```
  FCurrentPoint = ProjectOnSphere( ScreenPoint );
  LVector3 Direction = FCurrentPoint - FPrevPoint;
  LMatrix4 RotMatrix;
  RotMatrix.IdentityMatrix();
  float Shift = Direction.Length();
```

If the distance is non-zero, calculate and return a rotation matrix:

```
if ( Shift > Math::EPSILON )
{
  LVector3 Axis = FPrevPoint.Cross( FCurrentPoint );
  RotMatrix.RotateMatrixAxis( Shift * Speed, Axis );
}
FRotationDelta = RotMatrix;
return RotMatrix;
}
LMatrix4& GetRotationDelta()
{
  return FRotationDelta;
};
```

Get the current matrix:

```
virtual LMatrix4 GetRotationMatrix() const
{
  return FRotation * FRotationDelta;
}
static clVirtualTrackball* Create()
{
  return new clVirtualTrackball();
}
```

Reset the state of the trackball when a user touches the screen for the first time:

```
private:
  virtual void StartDragging( LVector2 ScreenPoint )
  {
    FRotation = FRotation * FRotationDelta;
    FCurrentPoint = ProjectOnSphere( ScreenPoint );
    FPrevPoint = FCurrentPoint;
    FRotationDelta.IdentityMatrix();
  }
```

Projection math goes here:

```
LVector3 ProjectOnSphere( LVector2 ScreenPoint )
{
  LVector3 Proj;
```

Convert normalized point coordinated to the `-1.0...1.0` range:

```
        Proj.x = 2.0f * ScreenPoint.x - 1.0f;
        Proj.y = -( 2.0f * ScreenPoint.y - 1.0f );
        Proj.z = 0.0f;
        float Length = Proj.Length();
        Length = ( Length < 1.0f ) ? Length : 1.0f;
        Proj.z = sqrtf( 1.001f - Length * Length );
        Proj.Normalize();
        return Proj;
    }
    LVector3 FCurrentPoint;
    LVector3 FPrevPoint;
    LMatrix4 FRotation;
    LMatrix4 FRotationDelta;
    bool FStarted;
};
```

The class is used in the `UpdateTrackball()` function, which is called from `OnDrawFrame()`:

```
    void UpdateTrackball( float Speed )
    {
      g_Trackball.DragTo(
        g_MouseState.FPos, Speed, g_MouseState.FPressed );
    }
    void OnDrawFrame()
    {
      UpdateTrackball( 10.0f );
      mat4 TrackballMtx = g_Trackball.GetRotationMatrix();
      Matrices.m_ViewMatrix =
        TrackballMtx * g_Camera.GetViewMatrix();
    }
```

This class enables you to rotate a 3D scene on a touchscreen and serves the purpose of debugging the scene on your device.

Summary

In this chapter, we learned how to build higher level scene graph abstractions on top of our platform-independent OpenGL wrapper. We can create scene objects with materials and light sources and render the scene with lighting and shading. In the next chapter, we will step away from rendering for a while—well, not entirely—and learn how to implement a game logic in C++.

Implementing Game Logic

9

The shortest description of this chapter contains only two words: state machines. Here, we will introduce a common approach to organize interactions between the gaming code and the user interface part of the application. We start with an implementation of the Boids algorithm and then proceed with the extension of our user interface implemented in the previous chapters.

Boids

In many game applications, you can usually see moving objects that collide, shoot, chase each other, can be touched or avoided by other objects, or produce similar behaviors. The visible complex behavior of objects can usually be broken down into a few simple states interoperating together. For example, in an arcade game an enemy randomly *wanders around* until it sees a player controlled character. After the encounter, it switches to the *chase* state, probably switching to *shoot* or *attack* states when in close proximity to the target. If an enemy unit perceives some disadvantage, it may *flee* from the player. The *chase* state in turn might not only target the enemy towards the player, but also avoid collisions with the environment. Each object can be differently animated or have different material while being in different states. Let's implement the chasing and wandering algorithms using an established approach invented by Craig Reynolds called *Flocking behaviors* or *Boids* (https://en.wikipedia.org/wiki/Boids). This method is used to create an impression of a semi-conscious flock or an intelligent swarm of some creatures. We use the State pattern (https://en.wikipedia.org/wiki/State_pattern) extensively throughout this chapter to define complex user interaction scenarios.

We consider only a two-dimensional game world and approximate each object, or a flockmate, as a circle with a velocity. Velocity, being a vector, has both magnitude and direction. Each object obeys three simple rules to calculate its desired velocity (http://www.red3d.com/cwr/boids):

1. *Avoidance*: Steer away to avoid crowding local flockmates.
2. *Cohesion*: Steer towards the average heading of local flockmates.
3. *Alignment*: Steer to move toward the average position of local flockmates.

Additional rules or behaviors that we implement are *ArriveTo* and *Wander* algorithms, which can be used as a basic debugging tool of our behavioral mechanics implementation.

The first rule, *Avoidance*, directs the velocity away from the obstacles as well as from the other flockmates, as shown in the following figure:

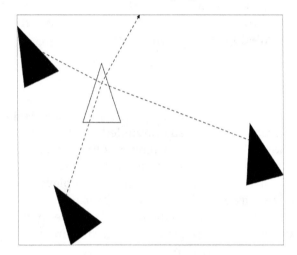

The second rule, *Cohesion*, steers towards the average heading of local flockmates, as shown in the following figure:

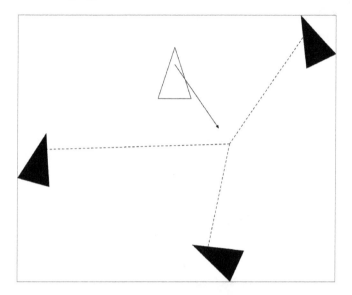

The third rule, *Alignment*, tries to adjust the calculated average velocity of nearby objects. It affects a group of flockmates in such a way that their movement direction soon becomes collinear and codirectional, as shown in the following figure:

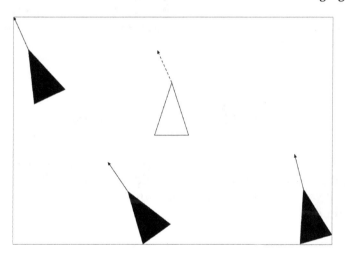

The rule, *ArriveTo*, sets the velocity direction to a predefined target point or region in space. By allowing the target to move in space, we can create some intricate behaviors.

Implementation of the flocking algorithms

To implement the above mentioned behaviors, we will consider the following class hierarchy:

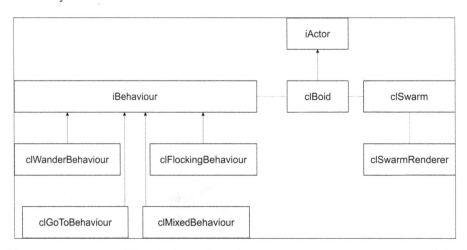

A single boid is represented by an instance of the `clBoid` class, which holds a pointer to an instance of `iBehaviour`. The `iBehavior` interface contains a single member function, `GetControl()`, which calculates the instant force acting on a boid. Since the force magnitude might depend on a boid's state, we pass a raw non-owning pointer to `clBoid` into `GetControl()`:

```
class iBehaviour: public iIntrusiveCounter
{
public:
  virtual vec2 GetControl( float dt, clBoid* Boid ) = 0;
};
```

Let's consider the `clBoid` class itself. It contains `m_Pos` and `m_Vel` fields, which hold the current position and velocity of the boid, respectively. These values are two-dimensional vectors, but the whole structure can be extended to 3D logic using three-component vectors:

```
class clBoid: public iActor
{
public:
  vec2 m_Pos;
  vec2 m_Vel;
```

The m_Angle field is the boid's instant orientation, and it is calculated from the m_Vel value. The m_MaxVel field contains the maximum velocity of the boid:

```
float m_Angle;
float m_MaxVel;
```

The m_Behaviour field holds a pointer to an iBehaviour instance, which calculates the control force of the desired behavior:

```
clPtr<iBehaviour> m_Behaviour;
```

Since a boid moves in a swarm and relies upon positions and velocities of the neighboring boids to adjust its velocity, we keep a non-owning pointer to the parent clSwarm object to avoid circular references between smart pointers:

```
clSwarm* m_Swarm;
```

The constructor of the class initializes default values and sets an empty behavior:

```
public:
  clBoid() :
  m_Pos(), m_Vel(),
  m_Angle(0.0f), m_MaxVel(1.0f),
  m_Behaviour(), m_Swarm(nullptr)
  {}
```

The only member function is Update(), which calculates a force acting on an object:

```
virtual void Update( float dt ) override
{
  if ( m_Behaviour )
  {
    vec2 Force = m_Behaviour->GetControl( dt, this );
```

After the force is calculated, it modifies the velocity according to Newton's law, $a = F/m$, and the Euler integration method (https://en.wikipedia.org/wiki/Euler_method). The mass of the boid is set to a constant value of 1.0. Feel free to introduce a varying parameter and observe how it changes the visual behavior of the swarm:

```
const float Mass = 1.0f;
vec2 Accel = Force / Mass;
m_Vel += Accel * dt;
}
```

To keep things visually plausible, we restrict the maximum possible velocity to the 0...m_MaxVel interval:

```
m_Vel = ClampVec2( m_Vel, m_MaxVel );
```

After the velocity is calculated, the position of the boid is updated:

```
m_Pos += m_Vel * dt;
```

Finally, the orientation of the boid should be evaluated as an angle between the X axis and the m_Vel vector. This value is used to render the boid on screen using a pointing arrow:

```
if ( m_Vel.SqrLength() > 0.0f )
{
  m_Angle = atan2( m_Vel.y, m_Vel.x );
}
  }
};
```

The simplest possible non-static behavior is random movement under the influence of abrupt random impulses. This is called the *Wandering* behavior and implemented in the clWanderBehaviour class. The GetControl() method calculates a vector with two random components within the -1..+1 range:

```
class clWanderBehaviour: public iBehaviour
{
public:
  virtual vec2 GetControl( float dt, clBoid* Boid ) override
  {
    return vec2( RandomFloat() * 2.0f - 1.0f,
      RandomFloat() * 2.0f - 1.0f );
  }
};
```

Another useful behavior of a boid is implemented in the clGoToBehaviour class. Given the target coordinates in the m_Target field, this behavior drives the controllable boid to that point. Once the boid is within the m_TargetRadius distance from m_Target, the movement is stopped:

```
class clGoToBehaviour: public iBehaviour
{
public:
```

The `m_Target` and `m_TargetRadius` fields define the location and radius of the target point:

```
vec2 m_Target;
float m_TargetRadius;
```

The `m_VelGain` and `m_PosGain` members hold two values, that define how fast a boid should brake once the target is reached, and how fast the boid accelerates proportionally to the distance to the target:

```
float m_VelGain;
float m_PosGain;
```

The constructor sets default values and non-zero gains:

```
clGoToBehaviour():
m_Target(),
m_TargetRadius(0.1f),
m_VelGain(0.05f),
m_PosGain(1.0f)
{}
```

The `GetControl()` routine calculates the difference between the boid position and the target. This difference is multiplied by `m_PosGain` and is used as the control force:

```
virtual vec2 GetControl( float dt, clBoid* Boid ) override
{
    auto Delta = m_Target - Boid->m_Pos;
```

If the boid is within the `m_TargetRadius` distance, we will return a zero as a value of the control force:

```
    if ( Delta.Length() < m_TargetRadius )
    {
        return vec2();
    }
```

A visually interesting braking effect can be achieved by replacing the preceding line with the line `return -m_VelGain * Boid->m_Vel / dt;`. The braking impulse is applied, which decreases the velocity by some fraction which results in a smooth exponential decay of the velocity. Visually, the boid stops smoothly near the target center.

The calculated impulse is returned at the end of the member function:

```
    return Delta * m_PosGain;
  }
};
```

Digression: helper routines

Here, we should describe a couple of functions used in control calculations.

In the preceding code, we used the `ClampVec2()` routine, which calculates the length of a vector `V`, compares this length to `MaxValue`, and returns either the same vector `V` or its clamped coaxial version of the `MaxValue` length:

```
inline vec2 ClampVec2(const vec2& V, float MaxValue)
{
  float L = V.Length();
  return (L > MaxValue) ? V.GetNormalized() * MaxValue : V;
}
```

Another bunch of methods include random number generation routines. The `RandomFloat()` method uses the C++11 standard library to generate uniformly distributed floating-point values in the 0…1 interval:

```
std::random_device rd;
std::mt19937 gen( rd() );
std::uniform_real_distribution<> dis( 0.0, 1.0 );
float RandomFloat()
{
  return static_cast<float>( dis( gen ) );
}
```

The `RandomVec2Range()` method uses the `RandomFloat()` function twice to return a vector with random components within a specified interval:

```
vec2 RandomVec2Range( const vec2& Min, const vec2& Max )
{
  return Min + vec2( RandomFloat() * ( Max - Min ).x,
    RandomFloat() * ( Max - Min ).y );
}
```

Collective behaviors

By now, we have defined only the `clWanderBehaviour` class. To implement flocking algorithms, we need to store information about all the boids at once. Such a collection is called a *Swarm* here. The `clSwarm` class holds a vector of `clBoid` objects and implements a number of routines used in boid control calculations:

```
class clSwarm: public iIntrusiveCounter
{
public:
  std::vector< clPtr<clBoid> > m_Boids;
  clSwarm() {}
```

For debugging and visual demonstration purposes, the `GenerateRandom()` method allocates a number of `clBoid` objects with random positions and zero velocities:

```
void GenerateRandom( size_t N )
{
  m_Boids.reserve( N );
  for ( size_t i = 0; i != N; i++ )
  {
    m_Boids.emplace_back( make_intrusive<clBoid>() );
```

By default, each boid has a *Wandering* behavior:

```
    m_Boids.back()->m_Behaviour =
      make_intrusive<clWanderBehaviour>();
    m_Boids.back()->m_Swarm = this;
```

Positions are random and are also kept within the -1..+1 range for each coordinate:

```
    m_Boids.back()->m_Pos =
      RandomVec2Range( vec2(-1, -1), vec2(1, 1) );
  }
}
```

The `Update()` method iterates the collection and updates every boid:

```
void Update( float dt )
{
  for ( auto& i : m_Boids )
  {
    i->Update( dt );
  }
}
```

The *Separation* or the *Avoidance* algorithm uses the sum of distances to other boids as the control force. The `clSwarm::CalculateSeparation()` method iterates the collection of boids and calculates the required sum:

```
vec2 CalculateSeparation( clBoid* B, float SafeDistance )
{
  vec2 Control;
  for ( auto& i : m_Boids)
  {
    if ( i.GetInternalPtr() != B )
    {
```

For each boid, except the one passed as the parameter, we calculate the position delta:

```
auto Delta = i->m_Pos - B->m_Pos;
```

If the distance is beyond the safety threshold, for example, if the boid is in close proximity with another boid, we add a negative delta to the control force:

```
      if ( Delta.Length() < SafeDistance )
      {
        Control += Delta;
      }
    }
  }
  return Control;
}
```

A similar routine is used in the *Cohesion* algorithm to calculate an average position of the neighboring boids:

```
vec2 CalculateAverageNeighboursPosition( clBoid* B )
{
  int N = static_cast<int>( m_Boids.size() );
```

We only sum the positions if there is more than one boid:

```
if ( N > 1 )
{
  vec2 Avg(0, 0);
```

A loop over the list of boids gives us the sum of the positions:

```
for ( auto& i : m_Boids )
{
  if ( i.GetInternalPtr() != B )
  {
    Avg += i->m_Pos;
  }
}
Avg *= 1.0f / (float)(N - 1);
return Avg;
}
```

In case of a single boid, we use its position. This way the control force in the *Cohesion* algorithm will be zero:

```
return B->m_Pos;
}
```

A similar procedure is applied to the velocities:

```
vec2 CalculateAverageNeighboursVelocity( clBoid* B )
{
  int N = (int)m_Boids.size();
  if (N > 1)
  {
    vec2 Avg(0, 0);
    for ( auto& i : m_Boids )
      if ( i.GetInternalPtr() != B )
        Avg += i->m_Vel;
        Avg *= 1.0f / (float)(N - 1);
    return Avg;
  }
  return B->m_Vel;
}
```

The utility method `SetSingleBehaviour()` sets the behavior of each and every boid in a swarm to the specified value:

```
void SetSingleBehaviour( const clPtr<iBehaviour>& B )
{
  for ( auto& i : m_Boids )
  {
    i->m_Behaviour = B;
  }
}
};
```

Now that we have the `clSwarm` class, we can finally implement the flocking behavior. The `clFlockingBehaviour` uses information about neighboring boids and calculates the control force with the classic Boids algorithms:

```
class clFlockingBehaviour : public iBehaviour
{
```

As usual, the constructor sets default parameters:

```
public:
  clFlockingBehaviour():
  m_AlignmentGain(0.1f),
  m_AvoidanceGain(2.0f),
  m_CohesionGain(0.1f),
  m_SafeDistance(0.5f),
  m_MaxValue(1.0f)
  {}
```

The `m_SafeDistance` field defines a distance at which the collision avoidance algorithm does not act:

```
float m_SafeDistance;
```

The next fields contain the weights for the influence of each flocking algorithm:

```
float m_AvoidanceGain;
float m_AlignmentGain;
float m_CohesionGain;
virtual vec2 GetControl(float dt, clBoid* Boid) override
{
  auto Swarm = Boid->m_Swarm;
```

The first step is *Separation* and *Avoidance*:

```
vec2 Sep = m_AvoidanceGain *
  Swarm->CalculateSeparation(Boid, m_SafeDistance);
```

The second step is *Alignment*:

```
auto AvgPos =
  Swarm->CalculateAverageNeighboursPosition(Boid);
vec2 Alignment = m_AlignmentGain *
  (AvgPos - Boid->m_Pos);
```

The third step is *Cohesion*. Steer to the average position of the neighbors:

```
auto AvgVel =
  Swarm->CalculateAverageNeighboursVelocity(Boid);
vec2 Cohesion = m_CohesionGain * (AvgVel - Boid->m_Vel);
```

Finally, we sum up all three values and keep the force magnitude below `m_MaxValue`:

```
  return ClampVec2(
    Sep + Alignment + Cohesion, m_MaxValue );
  }
};
```

The finishing touch for our behavior system is a class that implements a mixture of behaviors. The `clMixedBehaviour` class contains a vector of behaviors and respective weight factors that show how much of a behavior's control force is used in the resulting behavior:

```
class clMixedBehaviour : public iBehaviour
{
public:
  std::vector< clPtr<iBehaviour> > m_Behaviours;
  std::vector<float> m_Weights;
```

The `AddBehaviour()` member function adds a new weight factor and a behavior to the containers:

```
void AddBehaviour( float Weight, const clPtr<iBehaviour>& B )
{
  m_Weights.push_back( Weight );
  m_Behaviours.push_back( B );
}
```

As the name of the class suggests, the `GetControl()` routine calculates the control for each of the contained behaviors and sums all these control vectors multiplied by appropriate weights:

```
virtual vec2 GetControl(float dt, clBoid* Boid) override
{
  vec2 Control;
  for ( size_t i = 0; i < m_Behaviours.size(); i++)
  {
    Control += m_Weights[i] *
      m_Behaviours[i]->GetControl(dt, Boid);
```

```
        }
        return Control;
    }
};
```

As we can easily see, the `clFlockingBehaviour` class can be split up into *Avoidance*, *Cohesion*, and *Separation* parts. We decided not to complicate the structure of the book and implemented the flocking behavior as a single class. Feel free to experiment and mix these sub behaviors on your own.

Rendering the swarm

To use the developed swarm simulation system, we need to render individual boids. Since we already have an OpenGL 3D scene graph rendering system, we represent each boid with a triangular mesh and create scene nodes for them. Let's do this:

```
class clSwarmRenderer
{
private:
```

A single `clSceneNode` object in the `m_Root` field serves as the root scene node of the entire swarm:

```
    clPtr<clSceneNode> m_Root;
```

A pointer to the `clSwarm` object is kept to synchronize boid positions and angles with scene node transformations:

```
    clPtr<clSwarm> m_Swarm;
```

Scene nodes for each boid are stored in the `m_Boids` vector:

```
    std::vector< clPtr<clSceneNode> > m_Boids;
```

The constructor of the class creates a scene node for each boid in the swarm:

```
public:
    explicit clSwarmRenderer( const clPtr<clSwarm> Swarm )
    : m_Root( make_intrusive<clSceneNode>() )
    , m_Swarm( Swarm )
    {
        m_Boids.reserve( Swarm->m_Boids.size() );
        const float Size = 0.05f;
```

```
    for ( const auto& i : Swarm->m_Boids )
    {
      m_Boids.emplace_back(
        make_intrusive<clSceneNode>() );
```

Visually, the boid is a triangle, so we call `clGeomServ::CreateTriangle()` to create a vertex array with a single triangle:

```
    auto VA = clGeomServ::CreateTriangle(
      -0.5f * Size, Size, Size, 0.0f );
    auto GeometryNode =
      make_intrusive<clGeometryNode>( );
    GeometryNode->SetVertexAttribs( VA );
    m_Boids.back()->Add( GeometryNode );
```

Once a geometry node is initialized, we add it to `m_Root`:

```
    m_Root->Add( m_Boids.back() );
    }
    Update();
  }
```

At each frame, we calculate the transformation for each `clSceneNode` attached to the boid root node. The transformation consists of a translation into the boid's position followed by a rotation around the vertical z axis:

```
  void Update()
  {
    for ( size_t i = 0; i != m_Boids.size(); i++ )
    {
      float Angle = m_Swarm->m_Boids[i]->m_Angle;
      mat4 T = mat4::GetTranslateMatrix(
        vec3( m_Swarm->m_Boids[i]->m_Pos ) );
      mat4 R = mat4::GetRotateMatrixAxis( Angle,
        vec3( 0, 0, 1 ) );
      m_Boids[i]->SetLocalTransform( R * T );
    }
  }
  clPtr<clSceneNode> GetRootNode() const { return m_Root; }
};
```

All other scene management code is similar to that from the previous chapters:

Boids demonstration

The demo code in `1_Boids` uses a mixture of *GoTo* and *Flocking* behaviors to make a swarm of boids chase a user-specified target, and at the same time, create an illusion of the swarm-like movement.

We do not discuss the whole source of the application here and only underline the most important parts. The initialization of the demo starts with the creation of `clSwarm` filled with randomly positioned boids:

```
auto Swarm = make_intrusive<clSwarm>();
Swarm->GenerateRandom( 10 );
```

We set the same controller for all boids. The controller itself is a blend of `clFlockingBehaviour` and `clGoToBehavior` in the `g_Behaviour` object:

```
auto MixedControl = make_intrusive<clMixedBehaviour>();
MixedControl->AddBehaviour(0.5f,
  make_intrusive<clFlockingBehaviour>());
MixedControl->AddBehaviour(0.5f, g_Behaviour);
Swarm->SetSingleBehaviour(MixedControl);
```

The `g_Behaviour` instance holds coordinates of the target, which are initially set to `(1.0, 1.0)`:

```
g_Behaviour->m_TargetRadius = 0.1f;
g_Behaviour->m_Target = vec2( 1.0f );
g_Behaviour->m_PosGain = 0.1f;
```

The local `clSwarmRenderer` object is used at each frame of the rendering loop:

```
clSwarmRenderer SwarmRenderer( Swarm );
```

The demo uses touch input to change the coordinates of the target. When a touch occurs, we intersect the line passing through the touch point with the plane in which the boids reside. This intersection point is used as a new target point:

```
void OnTouch( int X, int Y, bool Touch )
{
  g_MouseState.FPos = g_Window->GetNormalizedPoint( X, Y );
  g_MouseState.FPressed = Touch;
  if ( !Touch )
  {
```

Once we know the touch has ended, we unproject the 2D mouse coordinates into the world space using perspective and view matrices:

```
vec3 Pos = Math::UnProjectPoint(
  vec3( g_MouseState.FPos ),
  Math::Perspective( 45.0f,
  g_Window->GetAspect(), 0.4f, 2000.0f ),
  g_Camera.GetViewMatrix() );
```

Using the camera view matrix, we calculate rotation and translation and use these values to intersect a ray from the mouse position with the z=0 plane:

```
mat4 CamRotation;
vec3 CamPosition;
DecomposeCameraTransformation( g_Camera.GetViewMatrix(),
  CamPosition, CamRotation );
vec3 isect;
bool R = IntersectRayToPlane(
  CamPosition, Pos - CamPosition,
  vec3( 0, 0, 1 ), 0, isect );
```

Once a 3D intersection point is constructed, it can be used as a 2D target of the *GoTo* behavior:

```
    g_Behaviour->m_Target = isect.ToVector2();
  }
}
```

At each iteration, we call the `Swarm::Update()` and `clSwarmRenderer::Update()` methods to update individual boid positions and velocities and to synchronize scene node transformations with the new data.

Now, go and run the `1_Boids` example to see for yourself.

The page-based user interface

Most parts of the previous chapters have laid down a foundation of a portable C++ application. Now, it is time to show you how to join more parts together. In *Chapter 7, Cross-platform UI and Input System*, we discussed how to create a simple custom user interface in C++ and respond to user input. In both cases, we only implemented a single fixed behavior without explaining how to switch to another one without writing spaghetti code. The first paragraphs of this chapter introduced the concept of *Behavior*, which we now apply to our user interface.

We call a single fullscreen state of the user interface as *Page*. Thus, every different screen of the application is represented by the `clGUIPage` class, which we annotate hereinafter.

Three main methods of `clUIPage` are `Render()`, `Update()`, and `OnTouch()`. The `Render()` method renders a complete page with all child views. `Update()` synchronizes the view with the application state. `OnTouch()` reacts to user input. The `clGUIPage` class is derived from `clUIView`, so there should not be any problems understanding this class.

The class contains two fields. The `FFallbackPage` field holds a pointer to another page, which is used as a return page, for example, when the back key is pressed on Android:

```
class clGUIPage: public clUIView
{
public:
```

The page we return to when the back key is pressed:

```
    clPtr<clGUIPage> FFallbackPage;
```

The non-owning pointer to the GUI object on this page came from:

```
    clGUI* FGUI;
public:
    clGUIPage(): FFallbackPage( nullptr ) {}
    virtual ~clGUIPage() {}
    virtual void Update( float DeltaTime ) {}
    virtual bool OnTouch( int x, int y, bool Pressed );
    virtual void Update( double Delta );
    virtual void SetActive();
```

The `OnActivation()` and `OnDeactivation()` methods are called when the GUI manager switches pages:

```
    virtual void OnActivation() {}
    virtual void OnDeactivation() {}
public:
    virtual bool OnKey( int Key, bool KeyState );
};
```

A list of pages is stored in the `clGUI` class. The `FActivePage` field indicates which page is currently visible. Events from user input are redirected to the active page:

```
class clGUI: public iObject
{
public:
  clGUI(): FActivePage( NULL ), FPages() {}
  virtual ~clGUI() {}
```

The `AddPage()` method sets a pointer to the parent GUI object and adds this page to the pages container:

```
void AddPage( const clPtr<clGUIPage>& P )
{
  P->FGUI = this;
  FPages.push_back( P );
}
```

The `SetActivePage()` method invokes some callbacks aside from actually setting the page as active. If the new page matches the currently active page, no action is performed:

```
void SetActivePage( const clPtr<clGUIPage>& Page )
{
  if ( Page == FActivePage ) { return; }
```

If we have a previously active page, we inform that page of switching to another page:

```
  if ( FActivePage )
  {
    FActivePage->OnDeactivation();
  }
```

If the new page is a non-null page, it is informed that it has been activated:

```
  if ( Page )
  {
    Page->OnActivation();
  }
  FActivePage = Page;
}
```

As we have mentioned before, each event is redirected to an active page stored in FActivePage:

```
void Update( float DeltaTime )
{
  if ( FActivePage )
  {
    FActivePage->Update( DeltaTime );
  }
}
void Render()
{
  if ( FActivePage )
  {
    FActivePage->Render();
  }
}
void OnKey( vec2 MousePos, int Key, bool KeyState )
{
  FMousePosition = MousePos;
  if ( FActivePage )
  {
    FActivePage->OnKey( Key, KeyState );
  }
}
void OnTouch( const LVector2& Pos, bool TouchState )
{
  if ( FActivePage )
  {
    FActivePage->OnTouch( Pos, TouchState );
  }
}
private:
  vec2 FMousePosition;
  clPtr<clGUIPage> FActivePage;
  std::vector< clPtr<clGUIPage> > FPages;
};
```

The implementation of the OnKey() method is only used in Windows or OSX. However, a similar logic can be applied to Android if we treat the back key as an analogue of the Esc key:

```
bool clGUIPage::OnKey( int Key, bool KeyState )
{
   if ( !KeyState && Key == LK_ESCAPE )
   {
```

If we have a non-null fallback page, we set it as active:

```
      if ( FFallbackPage )
      {
         FGUI->SetActivePage( FFallbackPage );
         return true;
      }
   }
   return false;
}
```

The implementation of SetActive() is put outside of the class declaration because it uses the then-undeclared clGUI class. This is used to remove the dependency from the header file:

```
void clGUIPage::SetActive()
{
   FGUI->SetActivePage( this );
}
```

Now, our mini GUI page mechanism is complete and can be used to handle user interface logic in an actual application.

Summary

In this chapter, we learned how to implement different behaviors of objects and use state machines as well as the design pattern to implement swarm logic. Let's proceed to the last chapter, so that we can combine many previous examples into a larger application.

10
Writing Asteroids Game

We will continue putting together the material from previous chapters. We will implement an Asteroids game with 3D graphics, shadows, particles, and sounds using techniques and code fragments introduced in the previous chapters. First, we will extend the previous material with a few more ideas, and then, we will move on and code a complete gaming application. Let's start with an onscreen joystick.

Creating an on-screen joystick

An on-screen joystick is based on multi-touch handling. Two structures contain descriptions of a single joystick button and an axis. The button is given an index and is specified by its color in the FColour field of the sBitmapButton structure. When a user taps the screen where the underlying pixel in the joystick mask has the color matching with one of the buttons, the clScreenJoystick class sets the pressed flag for the button:

```
struct sBitmapButton
{
  ivec4 FColour;
  int FIndex;
};
```

The sBitmapAxis structure represents a single stick of a joystick and contains two axes corresponding to vertical and horizontal directions. On the joystick mask bitmap, it is represented as a circular element centered at FPosition with the radius FRadius. The FAxis1 and FAxis2 indices specify which logical joystick axes are affected by this onscreen stick.

The `FColour` field is used to determine whether the user is touching this axis:

```
struct sBitmapAxis
{
  float FRadius;
  vec2 FPosition;
  int FAxis1, FAxis2;
  ivec4 FColour;
};
```

The `clScreenJoystick` class is declared as follows:

```
class ScreenJoystick: public iIntrusiveCounter
{
public:
  std::vector<sBitmapButton> FButtonDesc;
  std::vector<sBitmapAxis> FAxisDesc;
  std::vector<float> FAxisValue;
  std::vector<bool> FKeyValue;
```

This bitmap contains a colored mask with buttons:

```
  clPtr<clBitmap> FMaskBitmap;
public:
  ScreenJoystick()
  {}
```

Allocate button and axis state arrays:

```
void InitKeys()
{
  FKeyValue.resize( FButtonDesc.size() );
  if ( FKeyValue.size() > 0 )
  {
    for ( size_t j = 0 ; j < FKeyValue.size() ; j++ ) {
      FKeyValue[j] = false; }
  }
  FAxisValue.resize( FAxisDesc.size() * 2 );
  if ( FAxisValue.size() > 0 )
  {
    memset( &FAxisValue[0], 0,
      FAxisValue.size() * sizeof( float ) );
  }
}
```

Reset the state of the joystick buttons and axes:

```
void Restart()
{
  memset( &FPushedAxis[0], 0,
    sizeof( sBitmapAxis* ) * MAX_TOUCH_CONTACTS );
  memset( &FPushedButtons[0], 0,
    sizeof( sBitmapButton* ) * MAX_TOUCH_CONTACTS );
}
```

Check whether a button has been pressed:

```
bool IsPressed( int KeyIdx ) const
{
  return ( KeyIdx < 0 ||
    KeyIdx >= ( int )FKeyValue.size() ) ?
    false : FKeyValue[KeyIdx];
}
```

Get the current value of an axis:

```
float GetAxisValue( int AxisIdx ) const
{
  return ( ( AxisIdx < 0 ) ||
    AxisIdx >= ( int )FAxisValue.size() ) ?
    0.0f : FAxisValue[AxisIdx];
}
```

Button and axis setters are implemented in a similar way:

```
void SetKeyState( int KeyIdx, bool Pressed )
{
  if ( KeyIdx < 0 || KeyIdx >= ( int )FKeyValue.size() )
  { return; }
  FKeyValue[KeyIdx] = Pressed;
}
void SetAxisValue( int AxisIdx, float Val )
{
  if ( AxisIdx < 0 ||
    AxisIdx >= static_cast<int>( FAxisValue.size() ) )
  { return; }
  FAxisValue[AxisIdx] = Val;
}
```

Try to detect a button based on the color found in the joystick bitmap mask:

```
sBitmapButton* GetButtonForColour( const ivec4& Colour )
{
   for ( size_t k = 0 ; k < FButtonDesc.size(); k++ )
   if ( FButtonDesc[k].FColour == Colour )
   return &FButtonDesc[k];
   return nullptr;
}
```

The same logic applies to the axis detection:

```
sBitmapAxis* GetAxisForColour( const ivec4& Colour )
{
   for ( size_t k = 0 ; k < FAxisDesc.size(); k++ )
   {
      if ( FAxisDesc[k].FColour == Colour )
      { return &FAxisDesc[k]; }
   }
   return nullptr;
}
```

Currently pushed buttons and active axes are stored inside these member variables:

```
public:
   sBitmapButton* FPushedButtons[MAX_TOUCH_CONTACTS];
   sBitmapAxis* FPushedAxis[MAX_TOUCH_CONTACTS];
   void ReadAxis( sBitmapAxis* Axis, const vec2& Pos )
   {
      if ( !Axis ) { return; }
```

Read the value of an axis based on its center point and the touch point. The distance from the center point represents a value on a corresponding axis:

```
      float v1 = ( Axis->FPosition - Pos ).x / Axis->FRadius;
      float v2 = ( Pos - Axis->FPosition ).y / Axis->FRadius;
      this->SetAxisValue( Axis->FAxis1, v1 );
      this->SetAxisValue( Axis->FAxis2, v2 );
   }
};
```

The multi-touch handler is implemented in the following way:

```
void ScreenJoystick::HandleTouch(
  int ContactID, const vec2& Pos, bool Pressed, eMotionFlag Flag )
{
  if ( ContactID == L_MOTION_START )
  {
    for ( size_t i = 0; i != MAX_TOUCH_CONTACTS; i++ )
    {
      if ( FPushedButtons[i] )
      {
        this->SetKeyState(
          FPushedButtons[i]->FIndex, false );
        FPushedButtons[i] = nullptr;
      }
      if ( FPushedAxis[i] )
      {
        this->SetAxisValue(
          FPushedAxis[i]->FAxis1, 0.0f );
        this->SetAxisValue(
          FPushedAxis[i]->FAxis2, 0.0f );
        FPushedAxis[i] = nullptr;
      }
    }
    return;
  }
  if ( ContactID == L_MOTION_END )
  { return; }
  if ( ContactID < 0 || ContactID >= MAX_TOUCH_CONTACTS )
  { return; }
```

Clear all previous presses and axis states:

```
  if ( Flag == L_MOTION_DOWN || Flag == L_MOTION_MOVE )
  {
    int x = (int)(Pos.x * (float)FMaskBitmap->GetWidth());
    int y = (int)(Pos.y * (float)FMaskBitmap->GetHeight());
    ivec4 Colour = FMaskBitmap->GetPixel(x, y);
    sBitmapButton* Button = GetButtonForColour( Colour );
    sBitmapAxis* Axis = GetAxisForColour( Colour );
    if ( Button && Pressed )
    {
      // touchdown, set the key
      int Idx = Button->FIndex;
      this->SetKeyState( Idx, true );
```

Store the initial color of the button to track its movement later:

```
      FPushedButtons[ContactID] = Button;
   }
   if ( Axis && Pressed )
   {
     this->ReadAxis( Axis, Pos );
     FPushedAxis[ContactID] = Axis;
   }
  }
 }
}
```

To demonstrate the usage of the `clScreenJoystick` class, we modified the boids example from the previous chapter. The green box that denotes a target is controlled with the onscreen joystick.

Implementing the particle system

To make our game look shinier, we add yet another component to our rendering engine: a particle system. Particles move similarly to boids from the previous chapter, but vastly outnumber them and are not supposed to participate in complex interactions. Since individual particles are transparent, we need to take care of the rendering order and render particles after all solid objects within the frame have been rendered.

Each particle is treated as a point-like object when speaking of dynamics and rendered as a screen-aligned quadrilateral. A single particle does not exist forever and its initial lifetime `FLifeTime` and current time-to-live `FTTL` are stored. The `sParticle` structure contains `FPosition`, `FVelocity`, and `FAcceleration` fields describing its kinematic and dynamic properties. In addition to the physical properties, the `FRGBA` field contains a color of the particle and the `FSize` field describes its visual size. Let's put it this way:

```
struct sParticle
{
  sParticle(): FPosition(),
  FVelocity(),
  FAcceleration(),
  FLifeTime( 0.0f ),
  FTTL( 0.0f ),
  FRGBA( 1.0f, 1.0f, 1.0f, 1.0f ),
  FSize( 0.5f )
  {};
```

```
    LVector3      FPosition;              // current position
    LVector3      FVelocity;              // current velocity
    LVector3      FAcceleration;          // current acceleration
    float         FLifeTime;              // total life time
    float         FTTL;                   // time to live left
    LVector4      FRGBA;                  // overlay color
    float         FSize;                  // particle size
  };
```

For the simplicity of our implementation, we store particles as an **Array of Structures (AoS)** instead of **Structure of Arrays (SoA)**. The SoA approach is much more cache friendly and faster. If you are interested in how to implement a CPU-based particle system more efficiently, refer to this series of blog posts: http://www.bfilipek.com/2014/04/flexible-particle-system-start.html.

The private section of the clParticleSystem class contains a clVertexAttribs object with GPU-ready particle data, a container of sParticle instances, a material description for our rendering system, and the number of currently active particles:

```
class clParticleSystem: public iIntrusiveCounter
{
private:
  clPtr<clVertexAttribs> FVertices;
  std::vector<sParticle> FParticles;
  sMaterial FMaterial;
  int FCurrentMaxParticles;
```

The constructor preallocates vertices for an initial number of particles:

```
public:
  clParticleSystem(): FCurrentMaxParticles( 100 )
  {
    const int VerticesPerParticle = 6;
    FVertices = make_intrusive<clVertexAttribs>(
      VerticesPerParticle * FCurrentMaxParticles );
```

A special material class name is specified. Our rendering system will be aware of this material and will use proper shaders to render the particle system:

```
    FMaterial.m_MaterialClass = "Particle";
  }
  virtual void AddParticle( const sParticle& Particle )
  {
    FParticles.push_back( Particle );
```

If the number of particles exceeds current capacity of the vertex array, grow it using the coefficient 1.2. The optimal choice of growths coefficient here is subject to experiments and depends on emitters that are feeding the particle system:

```
if ( FCurrentMaxParticles <
   static_cast<int>( FParticles.size() ) )
{
   SetMaxParticles(int(FCurrentMaxParticles * 1.2f));
}
}
```

The SetMaxParticles() method adjusts the size of the FVertices vertex array to accommodate at least MaxParticles:

```
void SetMaxParticles( int MaxParticles );
```

We also need a bunch of getter member functions to access private fields of the class:

```
virtual std::vector<sParticle>& GetParticles()
{ return FParticles; }
virtual clPtr<clVertexAttribs> GetVertices() const
{ return FVertices; }
virtual const sMaterial& GetDefaultMaterial() const
{ return FMaterial; }
virtual sMaterial& GetDefaultMaterial() { return FMaterial; }
```

This is where everything happens. We will look into this method on the next page:

```
virtual void UpdateParticles( float DeltaSeconds );
};
```

The SetMaxParticles() method may seem simple, but it actually contains some useful code besides the trivial resizing of a container. To render particles, we use the technique called billboarding. For each particle, we create a screen-aligned quad consisting of two triangles. Texture coordinates of the quad corners are fixed, and for each particle, we fill the U and V values in the SetMaxParticles() method:

```
void clParticleSystem::SetMaxParticles( int MaxParticles )
{
   FCurrentMaxParticles = MaxParticles;
```

First, we will resize the FParticles array and FVertices object:

```
const int VerticesPerParticle = 6;
FParticles.reserve( FCurrentMaxParticles );
FVertices = make_intrusive<clVertexAttribs>
```

```
    ( VerticesPerParticle * MaxParticles );
  vec2* Vec = FVertices->FTexCoords.data();
```

Loop over the particles and assign six texture coordinate pairs to each vertex:

```
    for ( int i = 0; i != MaxParticles; ++i )
    {
      int IdxI = i * 6;
      Vec[IdxI + 0] = vec2( 0.0f, 0.0f );
      Vec[IdxI + 1] = vec2( 1.0f, 0.0f );
      Vec[IdxI + 2] = vec2( 1.0f, 1.0f );
      Vec[IdxI + 3] = vec2( 0.0f, 0.0f );
      Vec[IdxI + 4] = vec2( 1.0f, 1.0f );
      Vec[IdxI + 5] = vec2( 0.0f, 1.0f );
    }
  }
```

We synchronize particle coordinates, lifetime, and colors between FVertices and FParticles fields every frame. The UpdateParticles() method calculates new positions and velocities for each particle, and then updates the individual components of the FVertices object:

```
  void clParticleSystem::UpdateParticles( float DeltaSeconds )
  {
    vec3* Vec = FVertices->FVertices.data();
    vec3* Norm = FVertices->FNormals.data();
    vec4* RGB = FVertices->FColors.data();
    size_t NumParticles = FParticles.size();
    for ( size_t i = 0; i != NumParticles; ++i )
    {
      sParticle& P = FParticles[i];
      P.FTTL -= DeltaSeconds;
```

If the time-to-live of a particle is less than zero, we replace the particle with the last one in the array, so we can just efficiently pop the dead particle out of the container:

```
      if ( P.FTTL < 0 )
      {
        P = FParticles.back();
        FParticles.pop_back();
        NumParticles--;
        i--;
        continue;
      }
```

Using Newtonian physics and explicit Euler integrator, just as we have done for boids in the previous chapter, we recalculate the new velocity and position for each particle:

```
P.FVelocity += P.FAcceleration * DeltaSeconds;
P.FPosition += P.FVelocity * DeltaSeconds;
```

The time-to-live, total lifetime, and size of a particle are packed together into a `vec3` variable so that they can be stored in a vertex array:

```
LVector3 TTL = LVector3( P.FTTL, P.FLifeTime, P.FSize );
```

To keep formulas simple, we normalize the lifetime of the particle and use it in our color calculations:

```
float NormalizedTime = (P.FLifeTime-P.FTTL) / P.FLifeTime;
```

Depending on the normalized time, we calculate the color of the current particle. The `GetParticleBrightness()` function is described as follows:

```
vec4 Color = P.FRGBA *
    GetParticleBrightness( NormalizedTime );
```

Since each particle is represented by two triangles, we assign the same values to six consequential elements in the vertex array:

```
size_t IdxI = i * 6;
for ( int j = 0; j < 6; j++ )
{
  Vec [IdxI + j] = P.FPosition;
  Norm[IdxI + j] = TTL;
  RGB [IdxI + j] = Color;
}
}
```

After updating each particle, we adjust the number of vertices to be rendered in the vertex array to match the number of the currently active particles:

```
FVertices->SetActiveVertexCount
  ( 6 * static_cast<int>( FParticles.size() ) );
}
```

The `GetParticleBrightness()` function calculates a trapezoid-shaped function having a constant value of `1.0` for argument values from `0.1` to `0.9`. Visually, it means that at the beginning of the lifetime, particles fade in from zero to full visibility, then shine at a constant strength, and then linearly decay to zero:

```
inline float GetParticleBrightness( float NormalizedTime )
{
  const float Cutoff_Lo = 0.1f;
  const float Cutoff_Hi = 0.9f;
  if ( NormalizedTime < Cutoff_Lo )
  {
    return NormalizedTime / Cutoff_Lo;
  }
  if ( NormalizedTime > Cutoff_Hi )
  {
    return 1.0f - ( ( NormalizedTime - Cutoff_Hi ) /
      ( 1.0f - Cutoff_Hi ) );
  }
  return 1.0f;
}
```

At this point, we have only defined the class holding the particle instances. To integrate these new objects into our rendering system, we have to define a new kind of scene graph node, the `clParticleSystemNode` node. Before we can do this, a few words should be told about how the particles are emitted.

We introduce the `iParticleEmitter` interface, which declares a single pure virtual method, `EmitParticles()`, taking two parameters. The `DeltaTime` parameter is used to update the time counter and calculate new particle positions in the `PS` particle system:

```
class iParticleEmitter: public iIntrusiveCounter
{
public:
  iParticleEmitter():
  FColorMin( 0 ), FColorMax( 1 ),
  FSizeMin( 0.5f ), FSizeMax( 1.0f ),
  FVelMin( 0 ), FVelMax( 0 ),
  FMaxParticles( 1000 ),
  FEmissionRate( 100.0f ),
  FLifetimeMin( 1.0f ), FLifetimeMax( 60.0f ),
  FInvEmissionRate( 1.0f / FEmissionRate ),
  FAccumulatedTime( 0.0f )
  {}
  virtual void EmitParticles(
    const clPtr<clParticleSystem>& PS, float DeltaTime ) const = 0;
```

Fields of this class defines allowable ranges for each parameter of the particle. The limits for color, size, velocity, and lifetime are given by the variables with `Min` and `Max` postfixes. The `FEmissionRate` defines how many particles per second we are emitting and `FMaxParticles` gives the upper limit on the number of particles. The `FAccumulatedTime` field contains an approximate amount of time passed since the last particle system update:

```
public:
  vec4 FColorMin, FColorMax;
  float FSizeMin, FSizeMax;
  vec3 FVelMin, FVelMax;
  size_t FMaxParticles;
  float FEmissionRate;
  float FLifetimeMin, FLifetimeMax;
protected:
  float FInvEmissionRate;
  mutable float FAccumulatedTime;
};
```

The `EmitParticles()` method is overridden in two subclasses. The first of these subclasses is `clParticleEmitter_Box`, which emits particles in an axis-aligned box region:

```
class clParticleEmitter_Box: public iParticleEmitter
{
public:
  clParticleEmitter_Box(): FPosMin( 0 ), FPosMax( 1 ) {}
  virtual void EmitParticles( const clPtr<clParticleSystem>& PS,
    float DeltaTime ) const override
  {
    FAccumulatedTime += DeltaTime;
```

The following loop emits the required number of particles, one at a time. The position, velocity, color, time-to-live, and size are filled with uniform random variables:

```
while (
  FAccumulatedTime > 1.0f / FEmissionRate &&
  PS->GetParticles().size() < FMaxParticles )
{
  FAccumulatedTime -= 1.0f / FEmissionRate;
  sParticle P;
  P.FPosition = Math::RandomVector3InRange(
    FPosMin, FPosMax );
```

```
      P.FVelocity = Math::RandomVector3InRange(
        FVelMin, FVelMax );
      P.FAcceleration = LVector3( 0.0f );
      P.FTTL = Math::RandomInRange(
        FLifetimeMin, FLifetimeMax );
      P.FLifeTime = P.FTTL;
      P.FRGBA = Math::RandomVector4InRange(
        FColorMin, FColorMax );
      P.FRGBA.w = 1.0f;
      P.FSize = Math::RandomInRange(FSizeMin, FSizeMax);
      PS->AddParticle( P );
    }
  }
public:
  vec3 FPosMin, FPosMax;
};
```

This is one of the simplest emitters possible.

Using particle systems in the game

We also need a decent-looking explosion effect for our game. Particles emission in a combustive manner is implemented in the clParticleEmitter_Explosion class:

```
class clParticleEmitter_Explosion: public iParticleEmitter
{
public:
  clParticleEmitter_Explosion()
   : FEmitted( false ), FCenter( 0.0f )
   , FRadialVelocityMin( 0.1f ), FRadialVelocityMax( 1.0f )
   , FAcceleration( 0.0f )
   {}

  virtual void EmitParticles( const clPtr<clParticleSystem>& PS,
    float DeltaTime ) const override;
public:
  mutable bool FEmitted;
  vec3 FCenter;
  float FRadialVelocityMin, FRadialVelocityMax;
  vec3 FAcceleration;
};
```

The constructor sets the FEmitted fields to false. On the first invocation of EmitParticles(), this field is set to true and a fixed number of primary particles are emitted:

```
void clParticleEmitter_Explosion::EmitParticles(
  const clPtr<clParticleSystem>& PS, float DeltaTime ) const
{
  auto& Particles = PS->GetParticles();
  size_t OriginalSize = Particles.size();
```

The explosion effect adds a bunch of particles only once, but at each sequential EmitParticles() call secondary particles are created making trails, which follow the paths of primary particles. For each entity from the existing set of particles, an additional particle is created, so that the total number of particles is sustained within a budget set in the FMaxParticles variable:

```
for ( size_t i = 0; i != OriginalSize; i++ )
{
  if ( Particles[i].FRGBA.w > 0.99f &&
    Particles.size() < FMaxParticles )
  {
    sParticle P;
    P.FPosition = Particles[i].FPosition;
    P.FVelocity = Particles[i].FVelocity *
      Math::RandomVector3InRange( vec3(0.1f),
      vec3(1.0f) );
    P.FAcceleration = FAcceleration;
    P.FTTL = Particles[i].FTTL * 0.5f;
    P.FLifeTime = P.FTTL;
    P.FRGBA = Particles[i].FRGBA *
      Math::RandomVector4InRange( vec4(0.5f),
    vec4(0.9f) );
    P.FRGBA.w = 0.95f;
    P.FSize = Particles[i].FSize *
      Math::RandomInRange(0.1f, 0.9f);
    PS->AddParticle( P );
  }
}
```

Once we have created the explosion, this emitter will not work again:

```
if ( FEmitted ) return;
FEmitted = true;
```

The following loop creates a spray of particles with directions evenly distributed across a sphere:

```
for ( size_t i = 0; i != FEmissionRate; i++ )
{
  sParticle P;
```

Using two uniform random variables as angles, we calculate the uniform random direction:

```
float Theta = Math::RandomInRange( 0.0f, Math::TWOPI );
float Phi = Math::RandomInRange( 0.0f, Math::TWOPI );
float SinTheta = sin(Theta);
float x = SinTheta * cos(Phi);
float y = SinTheta * sin(Phi);
float z = cos(Theta);
```

Each particle starts at the center of explosion and the velocity coincides with the random direction calculated in the preceding code multiplied by a random magnitude:

```
P.FPosition = FCenter;
P.FVelocity = vec3( x, y, z ).GetNormalized() *
  Math::RandomInRange(
    FRadialVelocityMin, FRadialVelocityMax );
P.FAcceleration = FAcceleration;
```

Time-to-live, color, and size fields are filled with uniform random values:

```
P.FTTL = Math::RandomInRange(
  FLifetimeMin, FLifetimeMax );
P.FLifeTime = P.FTTL;
P.FRGBA = Math::RandomVector4InRange(
  FColorMin, FColorMax );
P.FRGBA.w = 1.0f;
P.FSize = Math::RandomInRange( FSizeMin, FSizeMax );
PS->AddParticle( P );
  }
}
```

Using particle systems inside a scene graph

Now, we are ready to declare and define the clParticleSystemNode class, which owns a clParticleSystem object, clGeometryNode with particle geometry, and a container of iParticleEmitter objects:

```
class clParticleSystemNode: public clMaterialNode
{
private:
  std::vector< clPtr<iParticleEmitter> > m_Emitters;
  clPtr<clParticleSystem> m_Particles;
  clPtr<clGeometryNode> m_ParticlesNode;
public:
  clParticleSystemNode();
  virtual void UpdateParticles( float DeltaSeconds );
  virtual clPtr<clParticleSystem> GetParticleSystem() const
  { return m_Particles; };
```

The following five methods provide access to the private particle emitters container:

```
  virtual void AddEmitter( const clPtr<iParticleEmitter>& E )
  { m_Emitters.push_back(E); }
  virtual void RemoveEmitter( const clPtr<iParticleEmitter>& E )
  {
    m_Emitters.erase( std::remove( m_Emitters.begin(),
      m_Emitters.end(), E ), m_Emitters.end() );
  }
  virtual clPtr<iParticleEmitter> GetEmitter( size_t i ) const
  { return m_Emitters[i]; }
  virtual void SetEmitter( size_t i,
    const clPtr<iParticleEmitter> E )
  { m_Emitters[i] = E; }
  virtual size_t GetTotalEmitters() const
  { return m_Emitters.size(); }
};
```

The constructor instantiates a particle system and all the necessary scene nodes:

```
clParticleSystemNode::clParticleSystemNode()
{
  m_Particles = make_intrusive<clParticleSystem>();
  size_t MaxParticles = 20000;
  for ( const auto& i : m_Emitters )
  {
```

```
    if (i->FMaxParticles > MaxParticles)
      MaxParticles = i->FMaxParticles;
  }
  m_Particles->SetMaxParticles(
    static_cast<int>(MaxParticles) );
```

Create a geometry node to store particles vertices:

```
  m_ParticlesNode = make_intrusive<clGeometryNode>();
  m_ParticlesNode->SetVertexAttribs(m_Particles->GetVertices());
  this->Add( m_ParticlesNode );
```

Get a material out of the particle system and apply it to the scene node:

```
  this->SetMaterial( m_Particles->GetDefaultMaterial() );
  UpdateParticles( 0.0f );
}
```

The `clParticleSystemNode::UpdateParticles()` method calls all the emitters, then invokes `clParticleSystem::UpdateParticles()` for `m_Particles`, and finally sends new particle vertex data to the rendering API using the `clGLVertexAray::CommitChanges()` call:

```
void clParticleSystemNode::UpdateParticles( float DeltaSeconds )
{
  for ( const auto& i : m_Emitters )
  {
    i->EmitParticles( m_Particles, DeltaSeconds );
  }
  m_Particles->UpdateParticles( DeltaSeconds );
  m_ParticlesNode->GetVA()->CommitChanges();
}
```

The rendering of `clVertexAttribs` that contains particle attributes requires new shaders to be written. Since particles represent a new type of geometry, we extend our `clMaterialSystem` class so that it can handle the particle material:

```
class clParticleMaterialSystem: public clMaterialSystem
{
public:
  clParticleMaterialSystem()
  {
    m_ParticleShaderPrograms[ ePass_Ambient ] =
      make_intrusive<clGLSLShaderProgram>( g_vShaderParticleStr,
        g_fShaderAmbientParticleStr );
```

```
    m_ParticleShaderPrograms[ ePass_Light ] =
      make_intrusive<clGLSLShaderProgram>( g_vShaderParticleStr,
        g_fShaderLightParticleStr );
    m_ParticleShaderPrograms[ ePass_Shadow ] =
      make_intrusive<clGLSLShaderProgram>(
        g_vShaderShadowParticleStr, g_fShaderShadowParticleStr );
}
```

The GetShaderProgramForPass() member function checks whether the material
class is *Particle* and picks a shader program from a new set of particle shader
programs. Otherwise, it falls back to the old clMaterialSystem implementation:

```
virtual clPtr<clGLSLShaderProgram> GetShaderProgramForPass(
  ePass Pass, const sMaterial& Mtl ) override
{
  if ( Mtl.m_MaterialClass == "Particle" )
  return m_ParticleShaderPrograms[ Pass ];
  return clMaterialSystem::GetShaderProgramForPass(
    Pass, Mtl );
}
```

The only new field in this class is a map that holds the new compiled shader
programs for each pass:

```
private:
  std::map<ePass, clPtr<clGLSLShaderProgram>>
  m_ParticleShaderPrograms;
};
```

Here goes the source code of all new shaders necessary to render particles. A vertex
shader is shared between all rendering passes and does the billboarding; this orients
particles to the camera:

```
static const char g_vShaderParticleStr[] =
R"(
  uniform mat4 in_ModelViewProjectionMatrix;
  uniform mat4 in_NormalMatrix;
  uniform mat4 in_ModelMatrix;
  uniform mat4 in_ModelViewMatrix;
  uniform mat4 in_ShadowMatrix;
  in vec4 in_Vertex;
  in vec2 in_TexCoord;
  in vec3 in_Normal;
  in vec4 in_Color;
  out vec2 v_Coords;
```

```
out vec3  v_Normal;
out vec3  v_WorldNormal;
out vec4  v_ProjectedVertex;
out vec4  v_ShadowMapCoord;
out vec3  v_Params;
out vec4  v_Color;
```

The same projected transform scale bias as for the default material:

```
mat4 GetProjScaleBiasMat()
{
  // transform from -1..1 to 0..1
  return mat4(
    0.5, 0.0, 0.0, 0.0,
    0.0, 0.5, 0.0, 0.0,
    0.0, 0.0, 0.5, 0.0,
    0.5, 0.5, 0.5, 1.0 );
}
void main()
{
```

Particles should be oriented so that they always face the camera. Let's construct a frame of reference where vectors X and Y are parallel to the screen:

```
vec3 X = vec3(
  in_ModelViewMatrix[0][0],
  in_ModelViewMatrix[1][0],
  in_ModelViewMatrix[2][0] );
vec3 Y = vec3(in_ModelViewMatrix[0][1],
  in_ModelViewMatrix[1][1],
  in_ModelViewMatrix[2][1] );
```

Fetch the particle size stored inside the normal's Z-component:

```
float SizeX = in_Normal.z;
float SizeY = in_Normal.z;
```

Use texture coordinates to calculate offsets from the center of the particle:

```
vec3 XOfs = ( 2.0 * (in_TexCoord.x-0.5) * SizeX ) * X;
vec3 YOfs = ( 2.0 * (in_TexCoord.y-0.5) * SizeY ) * Y;
vec3 Position = in_Vertex.xyz + XOfs + YOfs;
```

Transform the vertex position using the model-view-projection matrix:

```
vec4 TransformedPos =
  in_ModelViewProjectionMatrix * vec4( Position, 1.0 );
gl_Position = TransformedPos;
```

Pass on all other varying variables:

```
      v_Coords = in_TexCoord.xy;
      v_Normal = mat3(in_NormalMatrix) * in_Normal;
      v_WorldNormal =
        ( in_ModelMatrix * vec4( in_Normal, 0.0 ) ).xyz;
      v_ProjectedVertex = GetProjScaleBiasMat() *
        in_ModelViewProjectionMatrix * vec4(Position, 1.0);
      v_ShadowMapCoord = in_ShadowMatrix *
        in_ModelMatrix * vec4(Position, 1.0);
      v_Params = in_Normal;
      v_Color  = in_Color;
   }
 )";
```

Fragment shaders are more diverse and require a different one for ambient, shadow, and light passes to render particles properly. The following is the particle fragment shader for an ambient pass:

```
static const char g_fShaderAmbientParticleStr[] =
R"(
  in vec2 v_Coords;
  in vec3 v_Normal;
  in vec3 v_WorldNormal;
```

The value of `v_Params` comes from the `clParticleSystem::UpdateParticles()` function where TTL, lifetime, and size are packed together:

```
  in vec3 v_Params;
  in vec4 v_Color;
  out vec4 out_FragColor;
  uniform vec4 u_AmbientColor;
  uniform vec4 u_DiffuseColor;
  void main()
  {
    vec4 Color = v_Color * u_AmbientColor;
    float NormalizedTime =
      (v_Params.y-v_Params.x) / v_Params.y;
```

Compute transparency based on the distance to the particle center. This gives nice rounded particles without using any textures:

```
    float Falloff =
      1.0 - 2.0 * length(v_Coords-vec2(0.5, 0.5));
    if ( NormalizedTime < 0.1 )
    {
```

```
        Falloff *= NormalizedTime / 0.1;
      }
      else if ( NormalizedTime > 0.5 )
      {
        Falloff *= 1.0 - (NormalizedTime-0.5) / 0.5;
      }
      Color.w = Falloff;
      out_FragColor = Color;
    }
  )";
```

The fragment shader for a light pass looks as follows. It just discards the fragment, particles do not react on light sources:

```
static const char g_fShaderLightParticleStr[] =
R"(
  in vec2 v_Coords;
  in vec3 v_Normal;
  in vec3 v_WorldNormal;
  in vec4 v_ProjectedVertex;
  in vec4 v_ShadowMapCoord;
  in vec3 v_Params;
  in vec4 v_Color;
  out vec4 out_FragColor;
  void main()
  {
    discard;
  }
)";
```

A shadow map generation pass can be handled with the following fragment shader. Create a rounded shadow for each particle:

```
static const char g_fShaderShadowParticleStr[] =
R"(
  in vec2 v_Coords;
  in vec3 v_Params; /* TTL, LifeTime, Size */
  out vec4 out_FragColor;
  void main()
  {
    float NormalizedTime =
      (v_Params.y-v_Params.x) / v_Params.y;
```

The shadow grows for the first half of particle lifetime and shrinks to zero afterwards:

```
    float Falloff =
       ( NormalizedTime < 0.5 ) ?
       NormalizedTime : 1.0-NormalizedTime;
    if ( length(v_Coords-vec2(0.5, 0.5)) >
       0.5 * Falloff ) discard;
    out_FragColor = vec4( 1.0 );
  }
)";
```

A demonstration of particles rendering can be found in the `1_Particles` example. Here is a screenshot showing the running application:

At the beginning, we create an empty particle system node and pass it to the `GenerateExplosion()` function, which adds yet another explosion to the particle system. Here is how this is implemented:

```
void GenerateExplosion(
   const clPtr<clParticleSystemNode>& ParticleNode,
   const vec3& Pos )
{
```

The demo is intended to run on an Android device; do not spawn too many particles:

```
   if ( ParticleNode->GetParticleSystem()->GetParticles() .size()
      > 8000 ) return;
```

A palette for three different explosion types consists of blue, red, and green colors:

```
const vec4 Pal[] = {
  vec4(0.2f, 0.30f, 0.8f, 1.0f),
  vec4(0.7f, 0.25f, 0.3f, 1.0f),
  vec4(0.1f, 0.80f, 0.2f, 1.0f)
};
```

Just pick one tint randomly:

```
vec4 Color = Pal[ Math::RandomInRange(0, 3) ];
```

Create and set up the emitter object. You are strongly encouraged to play with these parameters:

```
auto Emitter = make_intrusive<clParticleEmitter_Explosion>();
Emitter->FCenter = Pos;
Emitter->FSizeMin = 0.02f;
Emitter->FSizeMax = 0.05f;
Emitter->FLifetimeMin = 0.1f;
Emitter->FLifetimeMax = 1.0f;
Emitter->FMaxParticles = 10000;
Emitter->FEmissionRate = 300;
Emitter->FRadialVelocityMin = 1.0f;
Emitter->FRadialVelocityMax = 2.0f;
Emitter->FColorMin = Color;
Emitter->FColorMax = Color;
Emitter->FAcceleration = vec3( 0.0f, 0.0f, -3.0f );
ParticleNode->AddEmitter( Emitter );
}
```

This function is invoked from the main loop:

```
while( g_Window && g_Window->HandleInput() )
{
  double NextSeconds = Env_GetSeconds();
  float DeltaTime = static_cast<float>(
    NextSeconds - Seconds );
  Seconds = NextSeconds;
  float SlowMotionCoef = 0.5f;
  if ( g_UpdateParticles )
  ParticleNode->UpdateParticles(
    SlowMotionCoef * DeltaTime );
```

Dice to decide whether we should add another explosion:

```
bool Add = Math::RandomInRange( 0, 100 ) > 50.0f;
```

Always add a new explosion if the particle system contains no active particles:

```
if ( !ParticleNode->GetParticleSystem()->
  GetParticles().size() || Add )
{
  GenerateExplosion( ParticleNode,
    Math::RandomVector3InRange(vec3(-1), vec3(+1)) );
}
OnDrawFrame();
g_Window->Swap();
}
```

Try building this demo for Android and running it on your device.

Asteroids game

Now, we have everything in place to deal with the actual game. Essentially, the game contains a lot of the previous examples stitched together to run in common and implement different aspects of the application. The glue that defines the game logic is in the class clGameManager, which is defined in Game.cpp and Game.h. The actual boid-like entities are implemented in Actors.cpp and Actors.h. Let's start with the base class iActor:

```
class iActor: public iIntrusiveCounter
{
public:
  iActor():
  m_Pos(0),
  m_Vel(0),
  m_Accel(0)
  {}
```

The major difference from all the previous examples is that there is no Render() method in this game framework. Instead, all entities should know how to attach to and detach from a scene graph. These methods are overridden in subclasses and vary between different types of actors:

```
virtual void AttachToScene(
  const clPtr<clSceneNode>& Scene ) = 0;
virtual void DetachFromScene(
  const clPtr<clSceneNode>& Scene ) = 0;
```

Some code is shared between all subclasses:

```
virtual void Update( float dt )
{
  m_Vel += m_Accel * dt;
  m_Pos += m_Vel * dt;
}
virtual float GetRadius() const
{
  return 0.1f;
}
public:
  vec3 m_Pos;
  vec3 m_Vel;
  vec3 m_Accel;
};
```

An asteroid game entity is an instance of the `clAsteroid` class, which is very simple:

```
class clAsteroid: public iActor
{
public:
  clAsteroid()
  : m_Angle( Math::RandomInRange( 0.0f, 1.0f ) )
  {}
  virtual void AttachToScene(
    const clPtr<clSceneNode>& Scene ) override;
  virtual void DetachFromScene(
    const clPtr<clSceneNode>& Scene ) override;
  virtual void Update( float dt ) override;
private:
  clPtr<clMaterialNode> m_Node;
  float m_Angle;
};
```

The implementation is almost trivial. Update the position and clamp it to the size of the game level:

```
void clAsteroid::Update( float dt )
{
  iActor::Update( dt );
  m_Angle += dt;
  m_Pos = g_Game->ClampToLevel( m_Pos );
  mat4 ScaleFix = mat4::GetScaleMatrix( vec3(0.002f ) );
  mat4 Pos = mat4::GetTranslateMatrix( m_Pos );
```

Asteroids always rotate around the (1,1,1) axis:

```
mat4 Rot = mat4::GetRotateMatrixAxis(
  m_Angle, vec3( 1, 1, 1 ) );
if ( m_Node )
  m_Node->SetLocalTransform( ScaleFix * Rot * Pos );
}
```

Attachment to the scene burns down to loading of an .obj file with an appropriate 3D model and setting up a material. A yellow one would be nice:

```
void clAsteroid::AttachToScene( const clPtr<clSceneNode>& Scene )
{
  if ( !m_Node )
  {
    auto Geometry = LoadOBJSceneNode(
      g_FS->CreateReader( "deimos.obj" ) );
    sMaterial Material;
    Material.m_Ambient = vec4( 0.5f, 0.5f, 0.0f, 1.0f );
    Material.m_Diffuse = vec4( 0.5f, 0.5f, 0.0f, 1.0f );
    m_Node = make_intrusive<clMaterialNode>();
    m_Node->SetMaterial( Material );
    m_Node->Add( Geometry );
  }
  Scene->Add( m_Node );
}
```

Detachment from the scene is simple:

```
void clAsteroid::DetachFromScene(
  const clPtr<clSceneNode>& Scene )
{
  Scene->Remove( m_Node );
}
```

The clRocket class represents a rocket fired from the space ship. Everything is similar to the implementation of clAsteroid except the Update() method:

```
void clRocket::Update(float dt)
{
  iActor::Update( dt );
  mat4 Pos = mat4::GetTranslateMatrix( m_Pos );
  if ( m_Node ) m_Node->SetLocalTransform( Pos );
```

If a rocket leaves the level area, kill it:

```
if ( !g_Game->IsInsideLevel( m_Pos ) )
{
  g_Game->Kill( this );
}
}
```

Explosions are implemented in the `clExplosion` class. The `clExplosion::AttachToScene()` method creates a particle system node with the emitter similar to the one in `GenerateExplosion()`. Nothing interesting there. However, the `Update()` method is slightly different:

```
void clExplosion::Update( float dt )
{
  iActor::Update( dt );
  mat4 ScaleFix = mat4::GetScaleMatrix( vec3(1.0f ) );
  mat4 Pos = mat4::GetTranslateMatrix(m_Pos);
  if ( m_Node )
  {
```

Particle system nodes require updating. Use a coefficient to make particles move slower:

```
    const float SlowMotionCoef = 0.1f;
    m_Node->SetLocalTransform( ScaleFix * Pos );
    m_Node->UpdateParticles( SlowMotionCoef * dt );
  }
```

Kill the explosion once all particles are gone:

```
  if ( !m_Node->GetParticleSystem()->GetParticles().size() )
  {
    g_Game->Kill( this );
  }
}
```

Last but not least, the `clSpaceShip` class represents a player-controllable entity. Again, the most interesting part is the `Update()` method, which handles the user controls:

```
void clSpaceShip::Update( float dt )
{
  iActor::Update( dt );
```

Ask the game manager if any control keys were pressed:

```
if ( g_Game->IsKeyPressed( SDLK_LEFT ) )
{
  m_Angle += dt;
}
if ( g_Game->IsKeyPressed( SDLK_RIGHT ) )
{
  m_Angle -= dt;
}
bool Accel = g_Game->IsKeyPressed( SDLK_UP );
bool Decel = g_Game->IsKeyPressed( SDLK_DOWN );
m_Accel = vec3( 0.0f );
if ( Accel )
{
  m_Accel = GetDirection();
}
if ( Decel )
{
  m_Accel += -GetDirection();
}
if ( g_Game->IsKeyPressed( SDLK_SPACE ) )
{
  Fire();
}
```

Make the ship warp between opposite sides of the level:

```
m_Pos = g_Game->ClampToLevel( m_Pos );
```

We do not want it to move very fast; here speed decay and clamping are implemented.

```
m_Vel *= 0.99f;
const float MaxVel = 1.1f;
if ( m_Vel.Length() > MaxVel )
  m_Vel = ( m_Vel / m_Vel.Length() ) * MaxVel;
```

A time counter is used to limit the rate of fire:

```
m_FireTime -= dt;
if ( m_FireTime < 0 ) m_FireTime = 0.0f;
```

Scale and rotate the 3D model to match the desired size and orientation:

```
mat4 ScaleFix = mat4::GetScaleMatrix( vec3(0.1f ) );
mat4 RotFix = mat4::GetRotateMatrixAxis(
    90.0f * Math::DTOR, vec3(0,0,1) );
mat4 Pos = mat4::GetTranslateMatrix(m_Pos);
mat4 Rot = mat4::GetRotateMatrixAxis( m_Angle, vec3(0,0,1) );
```

Apply the cumulative transformation:

```
if ( m_Node )
    m_Node->SetLocalTransform( ScaleFix * RotFix * Rot * Pos );
}
```

The `Fire` method does exactly what it appears to do. It launches a rocket and maintains the rate of fire:

```
void clSpaceShip::Fire()
{
    if ( m_FireTime > 0.0f ) return;
```

Try changing the weapon cool down time. One second is the default value:

```
const float FireCooldown = 1.0f;
m_FireTime = FireCooldown;
```

The actual rocket entity is added by the game manager:

```
g_Game->FireRocket(
    m_Pos, m_Vel * Math::RandomInRange( 1.1f, 1.5f ) +
    GetDirection() );
}
```

Those are all the entities existing in the game. Let's take a short glance at the `clGameManager` class, which rules them all:

```
class clGameManager: public iIntrusiveCounter
{
public:
    clGameManager();
```

Update the state of all objects and calculate collisions:

```
virtual void GenerateTicks();
```

Use the rendering technique to draw the game world:

```
virtual void Render();
virtual void OnKey( int Key, bool Pressed );
clPtr<clSceneNode> GetSceneRoot() const { return m_Scene; };
virtual bool IsKeyPressed( int Code );
```

There are two functions to create new entities; they are used in `clSpaceShip` and `CheckCollisions()`:

```
virtual void FireRocket( const vec3& Pos, const vec3& Vel );
virtual void AddExplosion( const vec3& Pos, const vec3& Dir );
```

A couple of high-level math functions to deal with positions of the entities:

```
virtual bool IsInsideLevel( const vec3& Pos );
virtual vec3 ClampToLevel( const vec3& Pos );
```

Kill a game actor, which can be an asteroid, an explosion, or a rocket. The space ship lives forever in our game. The `Kill()` method does not remove the actor immediately. Instead, it adds the actor to a container, which is later handled in the `PerformExecution()` method:

```
virtual void Kill( iActor* Actor );
```

The name says it all. Play an audio file in a fire-and-forget fashion:

```
  virtual void PlayAudioFile( const std::string& FileName );
private:
  void PerformExecution();
  void SpawnRandomAsteroids( size_t N );
  void CheckCollisions();
private:
  clPtr<clSceneNode> m_Scene;
  clPtr<clSpaceShip> m_SpaceShip;
  std::vector< clPtr<clAsteroid> > m_Asteroids;
  std::vector< clPtr<clRocket> > m_Rockets;
  std::vector< clPtr<clExplosion> > m_Explosions;
  std::unordered_map<int, bool> m_Keys;
  vec3 m_LevelMin;
  vec3 m_LevelMax;
  std::vector< iActor* > m_DeathRow;
  std::vector< clPtr<clAudioSource> > m_Sounds;
  // file name -> blob
  std::map< std::string, clPtr<clBlob> > m_SoundFiles;
};
```

The grand central dispatch of the gaming logic lays within the GenerateTicks() method:

```
void clGameManager::GenerateTicks()
{
  const float DeltaSeconds = 0.05f;
```

Update everything, check for collisions, and remove dead objects:

```
    for ( const auto& i: m_Asteroids ) i->Update( DeltaSeconds );
    for ( const auto& i: m_Rockets ) i->Update( DeltaSeconds );
    for ( const auto& i: m_Explosions ) i->Update( DeltaSeconds );
    m_SpaceShip->Update( DeltaSeconds );
    CheckCollisions();
    PerformExecution();
    for ( size_t i = 0; i != m_Sounds.size(); i++ )
    {
      if ( !m_Sounds[i]->IsPlaying() )
      {
```

Remove stopped audio sources one by one:

```
        g_Audio.UnRegisterSource(
          m_Sounds[i].GetInternalPtr() );
        m_Sounds[i]->Stop();
        m_Sounds[i] = m_Sounds.back();
        m_Sounds.pop_back();
        break;
      }
    }
}
```

Collision checking is done with a naive O(n^2) algorithm:

```
void clGameManager::CheckCollisions()
{
  for ( size_t i = 0; i != m_Rockets.size(); i++ )
  {
    for ( size_t j = 0; j != m_Asteroids.size(); j++ )
    {
      vec3 PosR = m_Rockets[i]->m_Pos;
      vec3 PosA = m_Asteroids[j]->m_Pos;
      float R = m_Asteroids[j]->GetRadius();
```

If a rocket is close enough to an asteroid, destroy both and add a huge explosion:

```
if ( (PosR-PosA).Length() < R )
{
  this->Kill(m_Rockets[i].GetInternalPtr());
  this->Kill(m_Asteroids[j].GetInternalPtr());
  AddExplosion( m_Asteroids[j]->m_Pos,
    m_Rockets[i]->m_Vel );
}
```

Executions are fast, but require a bit of C++ template magic:

```
void clGameManager::PerformExecution()
{
  for ( const auto& i : m_DeathRow )
  {
    i->DetachFromScene( m_Scene );
    Remove( m_Asteroids, i );
    Remove( m_Explosions, i );
    Remove( m_Rockets, i );
  }
  m_DeathRow.clear();
}
```

Here is the template code to handle heterogeneous entity containers:

```
template <typename Container, typename Entity>
void Remove( Container& c, Entity e )
{
  auto iter = std::remove_if(
    c.begin(), c.end(), [ e ](
      const typename Container::value_type& Ent )
    {
      return Ent == e;
    } );
  c.erase( iter, c.end() );
}
```

If you are a C++14 fan, you can definitely replace const typename Container::value_type& in the lambda parameter with const auto&, but our Visual Studio 2013 refused to compile the new code.

Whatever other functions are not mentioned here can be found in the `1_Asteroids` example. Build and run the code. It should look like this:

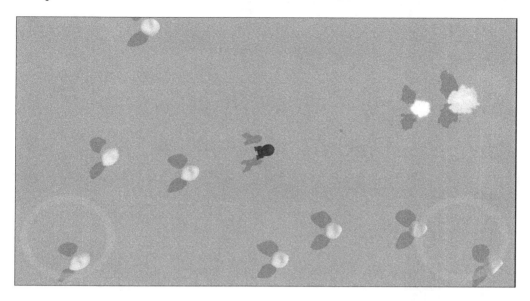

Summary

In this chapter, we summed up many techniques shown in the book and implemented a portable game application using the Android NDK. The essence of our tinkering with all that C++ code is the possibility to run our game unchanged on a desktop machine. This approach gives great opportunities in debugging large C++ mobile applications and faster iteration times when integrating new content into these apps. Furthermore, professional mobile development is never focused on one platform. With development practices like these, you can write C++ code that runs on many mobile platforms, including Android and iOS.

Index

Thank you for buying
Mastering Android NDK

About Packt Publishing

Packt, pronounced 'packed', published its first book, *Mastering phpMyAdmin for Effective MySQL Management*, in April 2004, and subsequently continued to specialize in publishing highly focused books on specific technologies and solutions.

Our books and publications share the experiences of your fellow IT professionals in adapting and customizing today's systems, applications, and frameworks. Our solution-based books give you the knowledge and power to customize the software and technologies you're using to get the job done. Packt books are more specific and less general than the IT books you have seen in the past. Our unique business model allows us to bring you more focused information, giving you more of what you need to know, and less of what you don't.

Packt is a modern yet unique publishing company that focuses on producing quality, cutting-edge books for communities of developers, administrators, and newbies alike. For more information, please visit our website at www.packtpub.com.

About Packt Open Source

In 2010, Packt launched two new brands, Packt Open Source and Packt Enterprise, in order to continue its focus on specialization. This book is part of the Packt Open Source brand, home to books published on software built around open source licenses, and offering information to anybody from advanced developers to budding web designers. The Open Source brand also runs Packt's Open Source Royalty Scheme, by which Packt gives a royalty to each open source project about whose software a book is sold.

Writing for Packt

We welcome all inquiries from people who are interested in authoring. Book proposals should be sent to author@packtpub.com. If your book idea is still at an early stage and you would like to discuss it first before writing a formal book proposal, then please contact us; one of our commissioning editors will get in touch with you.

We're not just looking for published authors; if you have strong technical skills but no writing experience, our experienced editors can help you develop a writing career, or simply get some additional reward for your expertise.

open source
community experience distilled

Android NDK Beginner's Guide
Second Edition

ISBN: 978-1-78398-964-5   Paperback: 494 pages

Discover the native side of Android and inject the power of C/C++ in your applications

1. Create high performance mobile applications with C/C++ and integrate with Java.

2. Exploit advanced Android features such as graphics, sound, input, and sensing.

3. Port and reuse your own or third-party libraries from the prolific C/C++ ecosystem.

Learning Android Intents

ISBN: 978-1-78328-963-9   Paperback: 318 pages

Explore and apply the power of intents in Android application development

1. Understand Android Intents to make application development quicker and easier.

2. Categorize and implement various kinds of Intents in your application.

3. Perform data manipulation within Android applications.

Please check **www.PacktPub.com** for information on our titles

open source
community experience distilled

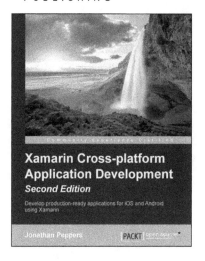

Xamarin Cross-platform Application Development
Second Edition

ISBN: 978-1-78439-788-3   Paperback: 298 pages

Develop production-ready applications for iOS and Android using Xamarin

1. Write native iOS and Android applications with Xamarin.iOS and Xamarin.Android respectively.

2. Learn strategies that allow you to share code between iOS and Android.

3. Design user interfaces that can be shared across Android, iOS, and Windows Phone using Xamarin.Forms.

Learning Android Application Testing

ISBN: 978-1-78439-533-9   Paperback: 274 pages

Improve your Android applications through intensive testing and debugging

1. Focus on Android instrumentation testing to ensure full application coverage.

2. Apply testing techniques and utilize tools to improve Android application development.

3. Build intensively tested and bug free Android applications.

Please check **www.PacktPub.com** for information on our titles